UNIVERSITY OF NORTH CAROLINA
STUDIES IN THE ROMANCE LANGUAGES AND LITERATURES

Number 50

I0029427

GIORDANO BRUNO'S
THE HEROIC FRENZIES

GIORDANO BRUNO'S
THE HEROIC FRENZIES

A TRANSLATION
WITH INTRODUCTION AND NOTES

BY

PAUL EUGENE MEMMO, JR.

CHAPEL HILL
THE UNIVERSITY OF NORTH CAROLINA PRESS

Second Printing

Copyright, 1966, by
The University of North Carolina Press

Reprinted 1966
by Permission
of the Original Publisher

GARRETT PUBLISHING COMPANY
103 West 49th Street
New York, N.Y. 10019

To the Three Theresas

PREFACE

For the primary source of my English translation of *De gli eroici furori* I have used Volume II of Giovanni Gentile's edition of Bruno's *Opere italiane.* Of all the editions of *De gli eroici furori* Gentile's is the most reliable. It takes into account the careful and accurate studies of Adolfo Wagner and Paul de Lagarde and often it modernizes much of Bruno's archaic spelling and punctuation while it remains completely faithful to Bruno's text in all other respects.

I have also consulted Paul-Henri Michel's French version of *Des fureurs héroïques,* published in 1954, and I have frequently made use of this felicitous translation in connection with many of Bruno's more obscure passages.

My purpose in this English version of *De gli eroici furori,* the first since William's translation of 1887, has been to provide a comprehensible reading which would at the same time be faithful to Bruno's original text. I must confess that for this purpose I have found Williams' translation of very little assistance, since to the floridity and looseness of Bruno's style it adds elements of confusion which are peculiar to the English.

For the occasional liberties I have taken with Bruno's language, I can only apologize on the ground of necessity. *De gli eroici furori* translated literally would become *Of the Heroic Furors,* a phrase which hardly carries Bruno's meaning into English. Williams translates it as *The Heroic Enthusiasts,* which seems equally baffling. What Bruno had in mind was obviously the Platonic furor, the divinely inspired frenzy of lovers. With a certain reluctance, therefore, I have departed from Bruno's title so far as to translate it as *The Heroic Frenzies,* which I hope will carry the necessary connotation and which has already had some acceptance. [1]

In order to render Bruno's style as faithfully as possible, I have maintained the structure of his elaborate and diffuse prose, insofar as it was possible to do so without obscuring the sense. I have italicized those words and phrases which were printed in spaced type in the edition of 1585. I have assumed that several of the mythological figures to which Bruno alludes are familiar and needed no annotation. I have, however, cited at some length those literary and philosophical allusions essential to an understanding of the work.

I should like to express my deepest gratitude to Professors Maurice Valency, Paul O. Kristeller and William Nelson of Columbia University for their advice and guidance in the preparation of this work. Above all, I owe special thanks to Professor Valency, my teacher, whose kindness beyond the call of duty led me to avenues I would perhaps not otherwise have explored in preparing this edition.

Moreover, I am grateful to those many colleagues and friends whose faith in the importance of this study has been of enduring inspiration. Accordingly, I should like to express my appreciation to Professors Marjorie Hope Nicolson, Joseph Satin, Robert McNulty, Francis X. Connolly, Grover J. Cronin, Charles J. Donohue, John F. Winter, Joseph P. Clancy and Joseph E. Grennen for their stimulating interest and warm encouragement which shall be remembered for years to come.

Last, and not the least of all, a word of thanks remains to be written to the many students of the Renaissance in Fordham University for their genuine enthusiasm, which taught me much over the years, even while I was teaching them.

PEM. Jr.

---

[1] DOROTHEA SINGER, *Bruno His Life and Thought* (New York, 1950), p. 125.

TABLE OF CONTENTS

INTRODUCTION

I

THE LONDON PERIOD AND *DE GLI EROICI FURORI*

The works of Giordano Bruno's London period (1583-1585), including *De gli eroici furori,* are those to which scholars have given most of their attention. These works contain a full development of the cosmological and ethical thought foreshadowed in the previous works of the Parisian period. [1] Their importance lies in the synthesis which Bruno achieved of the metaphysical thought that had preceded him in Europe since ancient times. *De gli eroici furori,* dedicated to Sir Philip Sidney, is the last of this group. [2]

London between 1583 and 1585, when Bruno lived there in the suite of the French Ambassador Michel de Castelnau was thought of as a haven for persecuted foreigners. Although Queen Elizabeth's court was renowned for the protection it offered to recalcitrant intellectuals of the Reformation, Bruno incurred the antagonism of scholars at Oxford. On one occasion he disputed in fifteen syllogisms against Dr. John Underhill, Rector of Lincoln College, in the presence of the Polish prince Laski. [3] Scholars have

---

[1] The Parisian period saw in 1582 the publication of *De umbris idearum,* a synthesis of Platonic and Plotinian philosophy, and the publication of Bruno's only play, *Il candelaio,* a satire of religious and social pedantry.

[2] For other editions of *De gli eroici furori* see ADOLFO WAGNER, *Opere di Giordano Bruno* (Lipsia, Weidmann, 1830), II; PAUL DE LAGARDE, *Opere italiane di Giordano Bruno* (Gottinga, 1888), II; GIOVANNI GENTILE, *Giordano Bruno, Opere italiane* (Bari, 1925), II; FRANCESCO FLORA, *Giordano Bruno, De gli eroici furori* (Torino, 1928); PAUL-HENRI MICHEL, *Des fureurs héroïques* (Paris, 1954).

[3] GIORDANO BRUNO, *Opere italiane,* 2 ed. Gentile (Bari, 1925), I, 101-102; LUDOVICO LIMENTANI, "Giordano Bruno a Oxford", *Civiltà moderna,*

speculated upon the nature of Bruno's dispute at Oxford, but no documentary evidence has been presented. Miss Frances Yates conjectures that Bruno opposed Oxford's repudiation of medieval metaphysics for a superficial preoccupation with mere grammar and style. [4]

Whatever the academic nature of the conflict, Bruno's erratic personality and often tactless egoism played no small part in aggravating his opponents. [5] Moreover, Bruno in the cosmological period vigorously attacked the traditional geocentrism of Aristotle and Ptolemy. One surmises that his substitution of a universe of innumerable solar systems unlimited in space and time, [6] while hardly controversial in our own day, did little to lessen the antagonism against him.

Bruno's cosmological works of this period consist of *La cena de le ceneri, De la causa, principio et uno*, and *De l'infinito universo et mondi*. The ethical works consist of *Spaccio de la bestia trionfante, Cabala del cavallo Pegaseo*, and *De gli eroici furori*. All the cosmological works are imprinted Venice, 1584. The first of the ethical works, *Spaccio de la bestia trionfante*, bears the imprint, Paris, 1584; the other two, Antonio Baio, Paris, 1585. However, the six works were actually printed in London by J. Charlewood. [7]

IX, No. 4-5 (July-Oct. 1937), p. 41, and ROBERT McNULTY, "Bruno at Oxford", *Renaissance News*, XIII, No. 4 (Winter, 1960), pp. 300-305.

[4] FRANCES YATES, "Giordano Bruno's conflict with Oxford", *Journal of Warburg Institute*, III (1939), 233.

[5] Bruno in a letter prefaced to some copies of *Recens et completa ars memorandi*, probably published in 1583, describes himself as one "whom only propagators of folly and hypocrites detest". See J. LEWIS McINTYRE, *Giordano Bruno* (London, 1903), p. 22.

[6] Bruno in *La cena de le ceneri* defends the Copernican thesis. However, he transcends the heliocentric theory of Copernicus and affirms a plurality of solar systems and a universe of innumerable stars unlimited in space and time. He conceives this as a view more consistent with religion than the traditional geocentric conception, for if God is to the world as cause is to effect, then from the infinity of God results the infinity of the world. See BRUNO, *La cena*, in *Opere italiane*, Gentile, I, 23-28; *De l'infinito*, in *Opere*, 340, 401-402. See also McNULTY, "Bruno at Oxford", 302-303.

[7] DOROTHEA SINGER, *Bruno His Life and Thought* (New York, 1950), p. 216.

The name of Antonio Baio on the two ethical works of 1585 is probably fictitious, for at his trial by the Inquisition in Venice, Bruno declared that his printers had advised him that the imprints of Venice and Paris would increase the sale of his books. [8]

The chief works of Bruno's Frankfort period (1590-1591) are *De Triplici minimo et mensura ad trium speculativarum scientiarum et multarum activarum artium principia libri v, De imaginum signorum et idearum compositione, De monade*, and *De innumerabilibus immenso et infigurabili seu de universo et mundis libri octo*. The four works were published by J. Wechel and P. Fischer in 1591. They contain in Lucretian Latin verse a further development of the philosophy and cosmology found in the works of the London period.

De gli eroici furori is the last of the three ethical works published in London and is the synthesis of Bruno's cosmological, ethical and poetic faith. The title evokes the subject matter: the ascension toward God and the return to the supreme unity of the soul through love. The work describes the progress of the soul in love as it mounts by degrees toward the supreme good. The description of this ascent, its states of progress and disillusionment before the final attainment of the ideal, is told in the form of a dialogue interspersed with sonnets.

For Bruno the term *eroici furori* has a very definite application. Plato in the *Phaedrus* refers to ἔρως as causing throbbing pulsations of the arteries, sleeplessness and pain, [9] and Greek physicians of the second century generally recognized ἔρως as a malady, the symptoms of which were pallor, excessive wakefulness, and a quickening of the pulse at the sight of the beloved object. Early Latin translators of the Greek medical texts generally associated ἔρως with the Latin *herus* or *heros*, hero, and formed the word, *hereos*. Thus Constantine's *Viaticus* translated ἔρως as *hereos*, to describe the love malady.

The medical treatises of the thirteenth and fourteenth centuries regularly include love sickness among the cerebral maladies. Accordingly, Bernardus Gordonius of Montpellier and John of

---

[8] *Ibid.*
[9] PLATO *Phaedrus* 251 D.

Gaddesden of Oxford both term the love malady, *amor heroycus, heroic love*, and treat its causes, signs, prognosis and cure. [10] Marsilio Ficino, in the Latin commentary upon Plato's *Symposium*, follows the tradition of the medieval medical treatises and examines it in the same way. [11]

Plato combines ἔρως with ἥρως and finds its origin in the older word, ἔρως :

> ... the name is not much altered, and signifies that they were born of love... the heroes... sprang either from the love of a God for a mortal woman, or of a mortal man for a goddess. Now if you think of the word in the old Attic, you will understand better that the name *heros* is only a slight alteration of *Eros* from whom the heroes sprang... [12]

The curious idea was thus established that *hereos*, the malady of love, was a disease particularly appropriate to heroes. For Bruno, therefore, the term "heroic" carries not only the nature of the lover's aspiration but also its nobility.

Bruno also follows Plato for his use of the word, *furori*, frenzies. Plato in the *Phaedrus* combines μαντίκη, prophecy, with μαν ική, frenzy, and defines both as the madness of prophets and poets:

> ...the ancient inventor of names would never have connected prophecy which foretells the future and is the noblest of the arts, with madness, or called them both by the same name, if they had deemed madness to be a disgrace or dishonor. They must have thought there was an inspired madness which was a noble thing; for the two words μαντίκη and μανική are really the same and the letter *t* is only a modern and tasteless insertion... [13]

---

[10] JOHN L. LOWES, "The Loveres Maladye of Hereos", *Modern Philology*, XI (1914), 419-546.

[11] MARSILIO FICINO, *In convivium*, in *Opere Platonis* (Paruo et Badio, 1518), VII, 3-11. The tradition will result in Robert Burton's extensive treatise in the seventeenth century on the love disease and its cure. See ROBERT BURTON, *The Anatomy of Melancholy*, in *Everyman* (London, 1948), III, pp. 3-257.

[12] PLATO *Cratylus* 398 C, D (Loeb, 1926), pp. 56-57.

[13] PLATO, 244 C (Loeb, 1926), pp. 466-467.

Plato describes four species of madness. The first is the prophetic madness; the second, the madness of the sacred ritual, which purges past and future evil; the third is poetic madness; and the fourth, the love madness, which is the greatest of heaven's blessings:

> ...Let him show in addition that love is not sent from heaven for the advantage of lover and beloved alike, and we will grant him the prize of victory. We, on our part, must prove that such madness is given by the gods for our greatest happiness... [14]

Marsilio Ficino in the *Commentarium in convivium* describes the four species of madness of the *Phaedrus*, but changes their order. Each species corresponds to one of the four degrees of the soul's ascension and is divinely inspired:

> ... Man by the divine madness is raised above his human nature, and ascends to God. Indeed this is a divine madness of the illumined spirit of reason: through it God from above draws up the soul fallen into the depth... Just as the soul descends by four degrees, so it must ascend by four. Thus there are four species of divine madness. The first then is the poetic madness inspired by the Muses. The second is the madness of the mysteries. The third is the prophetic madness. The fourth is the love passion. The poetic is (inspired) by the Muses, the mystical by Dionysos... the love madness by Venus... [15]

By the phrase, *eroici furori,* Bruno means to identify that species of intellectual aspiration which Plato describes in the *Symposium,* an aspiration which partakes of the highest nobility,

---

[14] *Ibid.,* 245 B (Loeb, 1926), pp. 468-469.
[15] FICINO, VII, 13-14:

> ... Divino furore super homini naturam erigitur, et in deum transit. Est autem furor divinus illustratio rationalis animae: per quam deus animam a superis delapsam ad infera, ab inferis ad supera retrahit... Sicut per quatuor descendit gradus, per quatuor ascendat necesse est... Quator ergo divini furoris sunt species. Primus quidem poeticus furor. Alter mysterialis. Tertius vaticinium. Amatorius affectus est quartus. Est autem poesis a musis, Mysterium a Dionysio, Vaticinium ab Apolline, Amore a Venere...

erotic in character, heroic in dignity, the fourth of the divine madnesses of the *Phaedrus*.

The dialogue of *De gli eroici furori* lacks the dramatic action and the pungent episodes of *La cena de le ceneri* and the *Spaccio*. The burden of the dialogue is carried simply by two persons. One of them expounds a doctrine. The other either approves it or demands clarification. When objections are raised, they are sometimes trivial and easily surmounted. Little attention is paid to the *mise en scène*, and the theme of conversation often begins and ends abruptly with such trite phrases as, "Let us begin" and "Let us return home".

However, Bruno constantly alternates with banal phrases those which are florid and extravagant. The hybrid use of language and mixture of learned and colloquial speech are reminiscent of Leone Ebreo's love treatise, *Dialoghi d'amore*, which appeared approximately in 1535. But despite the inconsistency of style, Bruno's dialogue, on the whole, is written in ardent poetic prose of genuine inspiration.

De gli eroici furori gives great importance to the sonnet. The philosophical dialogue serves as a gloss to the poetry in the tradition of the commentary upon love verses of the fourteenth and fifteenth centuries. Accordingly, the dialogue analyzes the sonnets according to levels of meaning which range beyond the literal to the allegorical, and perhaps anagogical. [16]

The grammatical commentary was a definite literary genre in the middle ages and late antiquity. Dante's commentary on the *Vita nuova* and *Convivio* are examples of the earlier form of grammatical commentary transferred to the vernacular literature. [17]

In the early fourteenth century Dino del Garbo wrote a Latin commentary on Guido Cavalcanti's *canzone, Donna mi prega*.

---

[16] See DANTE *Convivio* 2. i, for an explanation of the four levels of meaning. For Bruno's statement that he employs an allegorical level in *De gli eroici furori* see GIORDANO BRUNO, *Opere italiane*, Gentile, II, pp. 314, 315, 322-325; my translation, pp. 62-68, 73.

[17] For the statement that in the Middle Ages expositions of classical texts—of Cicero's rhetoric and Priscian's grammar, of Virgil, Ovid and Lucan—were the chief methods of teaching, see JOHN C. NELSON, *Giordano Bruno's Gli eroici furori and Renaissance Love Theory* (New York, 1958), pp. 21-22.

Shortly after, there appeared the supposed Egidio Colonna's vernacular commentary on the same *canzone,* the earliest formal vernacular commentary on vernacular verse.

Dante's *Vita nuova* and *Convivio* were the first Italian works in which a poet commented upon his own sonnets and *canzoni.* Late in the fifteenth century Lorenzo de Medici and Girolamo Benivieni explained their own sonnets, and Giovanni Pico della Mirandola wrote a commentary on Benivieni's *Canzone d'amore.*

In the same century appeared Marsilio Ficino's Latin commentary upon Plato's *Symposium,* which he then translated into the vernacular. This commentary was the first formulation of Platonic love in the Renaissance. After Ficino several Platonic love treatises appeared in the fifteenth and sixteenth centuries. Among them are Pietro Bembo's *Gli asolani* and Leone Ebreo's *Dialoghi d'amore.*

It is to the *trattati d'amore,* or the Platonic love treatises, that *De gli eroici furori,* in part, belongs. However, *De gli eroici furori* combines, for the first and the last time in Italian literature, the dialogue of the Platonic love treatise with the earlier tradition of prose commentary upon verses. [18]

The use of emblems as an introduction to many of the sonnets is an important device peculiar to *De gli eroici furori.* [19] Each emblem invokes a Latin motto. In the sixteenth century the emblem, like the poetic symbol, might express several levels of meaning from the literal to the allegorical and anagogical. [20] Accordingly, the philosophical dialogue serves as a gloss to the

---

[18] NELSON, pp. 3-25. Although both traditions employ the ladder of love in a way that is often indistinguishable, each employs it according to distinct philosophical origins. Therefore Dante, Cavalcanti and the prose commentaries upon their poetry do not, properly speaking, belong to the tradition of Platonic love in its original form. See PAUL O. KRISTELLER, *Renaissance Thought* (Harper, 1961), p. 64.

[19] Emblems introduce twenty-eight sonnets.

[20] Paolo Giovio and Girolamo Ruscelli make this point in their *Ragionamento.* See PAOLO GIOVIO and GIROLAMO RUSCELLI's, *Ragionamento di Mons. Paolo Giovio sopra i motti, et disegni d'arme, et d'amore, che communemente chiamono imprese. Con un discorso di Girolamo Ruscelli, intorno allo stesso soggetto* (Venezia, 1566), pp. 6, 7, 11, 175-179. The *Ragionamento* originally was published in 1555 under the title, *Dialogo dell'imprese militare e Amorosi.*

emblems as well as to the sonnets. Beneath its extraordinary complexity, therefore, the work, on the whole, has a fundamental coherence. [21]

---

[21] PAUL-HENRI MICHEL in *Giordano Bruno, Des fureurs héroïques*, p. lxxx, also holds this point of view. For the contrary view that *De gli eroici furori* suffers from obscurity and disorder see ANTONIO SARNO, "La genesi de gli eroici furori di Giordano Bruno", *Giornale critico della filosofia italiana*, II (1920), 158-172.

II

THE POETRY OF THE *STIL NOVISTI*

The particular basis of *De gli eroici furori* is the Petrarchistic sonnet sequence. Dante, by indicating in the *Convivio* the method of transforming a love song through allegory into a philosophical poem, had opened the way for every sort of abstruse interpretation of what were ostensibly songs of courtship and complaint. Moreover, from the time of Cavalcanti these songs written "in semblance of love" had combined elements of considerable philosophical interest.

The allegorical interpretation of love poetry had its origin in the method used by the medieval theologians to interpret the Scriptures. Origen and St. Augustine established the method of interpreting Scripture according to four levels of meaning. [1] It was further elaborated by St. Thomas Aquinas in the *Summa theologica*. According to the method, the words of Holy Scripture had first a literal or historical sense and upon this sense the other three are founded. [2] The Old Law is a figure of the New Law, and the New Law is a figure of future glory. As the New Law signifies the things men ought to do, that Law bears a moral sense. As the New Law signifies the journey of the soul to eternal bliss, it carries an anagogical sense. [3]

During the later Middle Ages the method of allegory was frequently used in poetry of a primarily philosophical nature. In

---

[1] St. Augustine *Confessions* xii. 31, 32.
[2] St. Thomas Aquinas *Summa theologica* I. i. 10.
[3] *Ibid.*

the twelfth century Bernardus Sylvestris' *De mundi universitate sive megacosmos et microcosmus* described the creation of man and the world by allegorical personifications in vogue since Prudentius had written the *Psychomachia.* In Neo-Plotinian fashion Bernardus recounts the creation in terms of the degrees of emanation from the eternal Mind to the lowest form of matter. Between the two extremes is man, an image of the whole world in whom the eternal and the temporal combine in perfect harmony. [4] Man's destiny is to rule over the earth and search its mysteries and after death to return in glory to heaven. [5] The νοῦς, Natura and Urania, all assume conventional allegorical personifications.

In the *Convivio* Dante combines the allegorical method of the medieval poets with the method used by the Church fathers to interpret Scripture. The tractates of the *Convivio* illustrate the application of the allegorical method to the conventional *canzone* of love. [6]

Dante divides the two primary kinds of meaning, the literal and the allegorical, into the four traditional senses of the theologians. The allegorical "hides itself under the mantle of these tales, and is a truth hidden under beauteous fiction. [7] The third meaning is the moral; and the fourth, the anagogical. As the theologians held, the three arise from the literal. But even in the literal sense the Scripture signifies things of man's eternal glory. When Israel was delivered out of Egypt, Judea became holy and free. This historic event, Dante tells us, allegorized the soul's sanctity and freedom when it is released from sin. [8]

Thus the honored lady of the courtly tradition acquires new metaphysical and theological significance. Through the operation of the allegorical method it became possible to equate her with the ideal of earthly beauty, or even with that absolute beauty

---

[4] BERNARDUS SYLVESTRIS, *De mundi universitate,* ed. Baruch and Wrobel (Innsbruck, Bibliotheca Philosophorum Mediae Aetatis, 1876), II, ix.

[5] *Ibid.,* II, iv. 31.

[6] DANTE *Convivio* 1. 1; see also MAURICE VALENCY, *In Praise of Love* (Macmillan, New York, 1958), p. 254.

[7] *Convivio* 2. 1.

[8] *Ibid.*

which is, in the concept of Guinizelli and of Dante, the ultimate goal of the lover of beauty.

In the scholastic world view the individual achieves his highest good by separating his spiritual from his corporeal nature rather than by attempting to bring about harmony between them. [9] Accordingly, the lover of the good does not consciously seek as his *Summum Bonum* a synthesis between his corporeal and his spiritual natures, as he will in *De gli eroici furori*. His true good lies rather in a renunciation of the flesh so that his soul may be set free for union with a knowable God. [10] Accordingly, in Dante's poetry the lady leads him to God through her beauty, and, presumably, releases him from the woe of sin. But it is her function as intermediary between God and man to vanish when her task is fulfilled. [11]

The lover in this phase of his progress finally comes to understand the beauty and goodness in the beloved were but reflections of the supreme Light. Since supreme goodness and light can exist perfectly only in the Supreme Being, he realizes his true goal from the very first has been God Himself. [12] When he has reached this last stage of understanding, his desire transcends into heaven and is filled with the love that "moves the sun and the other stars".

The basis of the *Vita nuova* was the poetry of the troubadours, but this poetry had suffered important transformations by the time of Dante. In the *Purgatorio* Dante has the poet, Bonagiunta da Lucca, speak of the *canzone*, "Donna ch'avete", as an example of the *dolce stil nuovo* that Dante and his fellow poets were practicing then in Florence. [13] This new style was based in part

---

[9] St. Thomas Aquinas, *Summa*, I. 76. 1. 84. 3; *De potentiis animae* I. 6.

[10] *Ibid.*

[11] Dante *Paradiso* 31.

[12] *Ibid.*, 33; see also Liborio Azzolina, *Il dolce stil nuovo* (Palermo, 1903), pp. 234-235.

[13] *Purgatorio*, 34, 52-57; see also G. Alfredo Cesareo, "Amor mi spira", in *Miscellanea di studi critici*, Arturo Graf (Bergamo, 1903), pp. 515-525.

upon the work of Guido Guinizelli, [14] whom Dante hailed in the *Purgatorio* as his "father".

Dante in the *Vita nuova* describes his progress toward the absolute from the first intuition of the ideal in the beloved. Before he achieves ultimate victory, he must endure repeatedly rejection by his lady because of his unworthiness. Dante makes her frequent denials the spur to a higher stage of spiritual progress. However, full realization of Beatrice's excellence is not possible while she remains alive.

Departing from the tradition of the troubadours, Dante makes the death of Beatrice the central event in the *Vita nuova*. [15] It is the death of Beatrice that finally releases her love for the highest stage 'of Dante's spiritual evolution. Accordingly, the loss of Beatrice to this world spurs him to renounce the life of the senses for the new life of intellectual and divine contemplation. The new life, then, begins in death.

The final sonnet of the *Vita nuova* describes the poet's intuition of the significance of Beatrice, whom he sees "beyond the sphere that circles widest". [16] In the *Convivio* Dante explains this sphere as the *primum mobile* that lies between the world of planetary spheres and the angelic spheres of the empyrean heaven. [17] The poet conceives Beatrice as an angelic intelligence receiving homage from the angels. In this way Dante anticipates the vision of Beatrice in the closing cantos of the *Paradiso* as she takes her place in the third rank of the mystical rose, while he ascends toward final contemplation of the supreme Light. [18] The

---

[14] Guido Guinizelli's *canzone*, "Al cor gentil ripara sempre Amore", transforms the *domina* of the troubadours into the *donna angelicata* of the *stilnovisti*. See CHARLES SINGLETON, *Essay on the Vita nuova* (Cambridge, Mass., 1949), pp. 1-50.

[15] Dante's contemplation of Beatrice's death and of his own mortality in the "Donna pietosa" occurs in nine lines designed to occupy the precise center of the *canzone;* see DANTE *Vita nuova* 23. The *Vita nuova* of which this *canzone* is the center, is composed of three decads of poems and a final sonnet: three and one, a trinity. For Dante's use of mystic numerology see MAURICE VALENCY, *In Praise of Love*, pp. 264-266; see also CHARLES S. SINGLETON, *An Essay on the Vita nuova*, pp. 1-50.

[16] DANTE *Vita nuova* 42.

[17] *Convivio* 2. 4.

[18] *Paradiso* 31, 60-74. For the suggestion that the "donna angelicata"

conceit of the lady who transcends the *domina* of the troubadours to receive homage of the cosmic spheres is of the essence of the new style. The new poetry often strikes one as audaciously learned, and even sophisticated. But it has a cosmic sweep that makes one marvel.

In the *Convivio* Dante continues his progress toward the *summum bonum* by means of a new love whom he describes as *Donna Sapienza*. [19] This is Philosophy, and from her place in the third heaven of thrones she illuminates the poet's intellect so that he may contemplate the cosmic order beyond the sensible world. [20] Only after he has absorbed the teaching of philosophy may he ascend through the beauty of Beatrice to the final vision of the eternal Light, whose divine image is reflected in the whole cosmic creation. [21] Accordingly, Dante assigns the lady a role related peculiarly to the traditionally geocentric view of the universe.

Petrarch's *Rime* are reminiscent of the lyrics of the Provencal and Italian troubadours, as well as of the lyricism of the school of Dante. Petrarch emphasizes particularly the lover's personal plight, and his verses have a more intimate subjectivity than we are accustomed to among the *stilnovisti*. To a certain extent the antithetical style that characterizes much of his verse is the result of the inner conflict, which forms the principal subject of his *canzonieri*. [22]

As Petrarch intimates in his *Secretum*, the lover's contrary states reflect the war between the sensuous and the spiritual elements in his nature, which attends the gradual ascent of desire to its highest object. In the *Secretum* St. Augustine tells Petrarch that if he renounces earthly love and glory, it is not because he despises

---

belongs essentially to the universal intelligence of Averroës in which the lover's intellect participates only on occasion, see KARL VOSSLER, *Die philosophischen Grundlagen zum süssen neuen Stil* (Heidelberg, 1904), pp. 79-81.

[19] *Convivio* 2. 16; see also Azzolina, pp. 170-173, 185-189.

[20] See *Convivio* 2. 6 and 14, for the correspondence existing among the angelic intelligences, planetary spheres and the order of the sciences, and the influence of this threefold correspondence upon the individual soul.

[21] *Paradiso* 33, 48-94.

[22] For the description of the contrary passions of the lover see PETRARCH *Rime* 134.

them, but because he despairs of attaining them. [23] For this reason, Augustine suggests, it is necessary to love the creature only in order to arrive at the divine love of the Creator. [24]

With Cavalcanti the lover's internal strife is the battle between the ego and the object of desire, the tyrranical image of beauty that is lodged in the lover's heart. But after Petrarch the *battaglia d'amore* resolved itself conventionally in the conflict of the higher and the lower soul and the higher and the lower loves. In the next centuries Lorenzo de Medici's *Commento sopra alcuni dei suoi sonnetti*, Girolamo Benivieni's *Canzone d'amore*, Ficino's commentary upon Plato's *Symposium*, and Michelangelo's sonnets, all reflect some aspect of the lover's conflict in these terms. [25] All these works develop the course of one involved in the scale of love, the summit of which is the absolute. Among the innumerable lyric sequences that were elaborated on this theme in the sixteenth century, the sonnet sequence of Luigi Tansillo has an important place. Tansillo's sequence is especially important for an understanding of Giordano Bruno's *De gli eroici furori*, since Bruno employed several sonnets from Tansillo's *Poesie liriche* as the basis for his commentary in *De gli eroici furori*. [26]

Luigi Tansillo [27] belongs to the group of Neapolitan Petrarchists which also includes Iacopo Sannazaro, Benedetto Cariteo, and Antonio Epicuro. In 1535 while enjoying the patronage of Pietro de Toledo, Spanish viceroy of Naples, Tansillo composed his *Poesie liriche* to a lady believed to have been the wife of the Marchese del Vasto, Marie of Aragon, who rejected his love. [28]

---

[23] PETRARCH *Secretum* 2.

[24] *Ibid.*, 3.

[25] For an analogous use of the *battaglia d'amore* in England in the sixteenth century see SIR THOMAS WYATT, son., "I find no peace, and all my war is done"; SIR PHILIP SIDNEY, in *Astrophel and Stella*, son., "It is most true that eyes are formed to serve"; see also EDMUND SPENSER, *Amoretti*, 3, 8, 18, 34.

[26] See BRUNO, *Opere italiane*, 2 ed. Gentile, II, 342, 363, 369, 400, 412, 490, 494, 512, 518, for the origin of four of the sonnets of *De gli eroici furori* in LUIGI TANSILLO's *Poesie liriche*.

[27] Luigi Tansillo (1510-1568), b. Venosa of Nolan ancestry, served the Spanish viceroy, Pietro de Toledo, in the war against the Turks in 1537.

[28] LUIGI TANSILLO, *Poesie liriche*, ed. Fiorentino (Napoli, 1882), pp. xliv, xlv.

Tansillo's sequence is in no way unconventional for the period. The theme, as always, is disprized love, and in pursuit of the merciless fair, the lover suffers the usual torments, the warring passions of love and hate, desire and jealousy. For Tansillo, however, *gelosia* has a special meaning. It is the pain suffered through the deprivation of the beloved object.[29] He describes this pleasurable pain in terms entirely reminiscent of Petrarch, his spiritual ancestor:

> Dear, gentle and revered wound of that sweet
> dart, which love ever chooses, lofty, gracious, and
> precious ardor, which makes the soul toss in ever
> burning delight,
>
> what virtue of herb, or force of magic art,
> will ever release you from the center of my
> heart, if the fresh onslaught which strikes
> there at every hour, delights me the more it
> torments me?[30]

Characteristics of Tansillo's sonnets is an ebullient use of imagery and the sort of hyperbole that is common among the later Petrarchists. In time, Tansillo's lyrics will result in the "baroque" conceit of the following decades.[31] Cupid's arrow reappears in new and elaborated combinations:

---

[29] ANDREAS CAPPELLANUS, *De amore*, trans. John J. Barry (New York, 1941), pp. 45, 46, 102, 158, 185; see also LORENZO DE MEDICI, *Commento sopra alcuni dei suoi sonnetti*, in *Opere* (Bari, 1913), II, p. 93.

[30] Tansillo, p. 15:

> Cara, soave, ed onorata piaga
> Del piú bel dardo, che mai scelse Amore,
> Alto, leggiadro, e precioso ardore,
> Che gir fai l'alma di semp' arder vaga,
>
> Qual virtú d'erbe, o forza d'arte maga,
> Ti torrà mai dal centro del mio core,
> Si chi vi porge ognor fresco vigore,
> Quanto piú mi tormenta, piú m'appaga?

[31] I employ the term "baroque" in the sense of the hyperbolic image which appears frequently to dazzle the reader for its own sake; see, for example, GIOVANNI MARINO, *Adone* 8, 1-23, and RICHARD CRASHAW in England, "The Tear".

Eyes, flames and bow of my lord, twofold fire
in the soul, and arrows in the heart, because
the languishing is sweet to me, and the fire is dear. [32]

In Tansillo's *Due pellegrini* two young men, Alcinio and Filauto, have experienced the first effects of Cupid's dart when the poem
begins. Each laments his lady's loss, and each seeks death. Al-
cino's lady is still alive but seeks the love of another, while
Filauto's lady has been dead for seven years. This situation
initiates a *dubbio d'amore*. For Tansillo each lover suffers from
an essentially different kind of blindness and jealousy. Alcinio
suffers from the blindness and jealousy caused by the loss of
a living lady. [33] Filauto suffers from the blindness and jealousy
caused by the loss of a lady who is dead. [34] None can say which
is the greater suffering. The *dubbio* is solved for both lovers when
Filauto's lady suddenly appears in a vision. She scolds the lovers
for despairing and advises each according to the progress he has
made in the scale of love. In the manner of the *stil novisti* Tansillo
makes her acknowledge Filauto's martyrdom in the service of
chaste love and assures him of heaven's solicitude:

...Inviolable faith, chaste love, the highest
good, tears, martyrdom, friends too dear to your
heart; and your lofty desire... had so much
strength in the third Heaven, that I am sent down
here to prevent your implacable and wilful death... [35]

---

[32] Tansillo, p. 15:

Occhi, del mio Signor facelle ed arco,
Doppiate fiammc a l'alma, e strali al petto,
Poi che 'l languir m'è dolce, e l'ardor caro.

[33] Luigi Tansillo, *Due pellegrini* (Napoli, 1631), pp. 5, 6.
[34] *Ibid.*, p. 9.
[35] *Ibid.*, p. 38:

... L'inviolabil fede, il casto amore,
L'alta bontà, le lagrime, il martire
Amici troppo cari del tuo core;
E'l tuo alto desire...
Hebber nel terzo Ciel tanto vigore,
Che mi trasser quì per impedire
La tua spietata, e volontaria morte...

Her advice is to some extent reminiscent of that given by Beatrice to Dante in the *Purgatorio*. [36] She warns Alcinio that he has yet to renounce the earthly lady and that he must learn to adore her celestial form. Laura in a vision after her death had given Petrarch similar advice. [37] In this manner, through the renunciation of the carnal appetite, he may find a new way of life, "prendendo della vita un nuovo stile":

> ...As the serpent casts off the enfeebled skin,
> and reveals the new to a grateful April, so will you
> alter your old desire. Choosing a new manner of life,
> no desire, thought or pain for any vile thing shall ever
> live in your heart... [38]

The lady's advice to Filauto implies that he has progressed further than Alcinio upon the ladder of love. She explains how Filauto's pure and faithful love since her death seven years before has summoned her from her place among the thrones in the third heaven, so that she might descend to quicken his soul to beatitude. [39] Filauto's blindness is ultimately cured when he enjoys a final vision of his lady reascending to the choir of angels. [40]

The new style, therefore, had already seen much progress when Bruno wrote *De gli eroici furori*, and it was natural that he might borrow from it. However, he borrowed only those elements that served the demands of his own philosophy.

---

[36] DANTE *Purgatorio* **30**.

[37] PETRARCH *Rime* 285.

[38] TANSILLO, *Due pellegrini*, p. 45:

> ... Come il serpente l'invecchiata spoglia
> Gitta, e la nova scopre al grato Aprile,
> Così tu congerai l'antica voglia.
> Prendendo della vita un nuovo stile,
> Ne giammai fiamma, ne pensier ne doglia
> Vivran dentro il tuo cor di cosa vile...

[39] *Ibid.*, p. 38. Tansillo follows Dante and associates the lady with the mystic numbers three and seven.

[40] *Ibid.*, p. 46.

III

THE SONNET SEQUENCE OF *DE GLI EROICI FURORI*

Like the earlier sequences of Dante and Petrarch, Bruno's sonnets narrate the progress of the soul toward the absolute from the first intuition of the beauty of the ideal. As Bruno tells it, this progress is not an easy matter. Before victory crowns its efforts, the soul must endure repeated reversals and moments of acute despair.

Bruno's sonnets describe the states of the lover's soul as he alternates between progress and regression in the course of his arduous journey toward the supreme good. After each of its setbacks along the upward path to the ideal, the soul achieves a new degree of inner balance. But this balance is momentary. As further progress is made, there is further internal conflict, which is again resolved in a new equilibrium more perfect than before and on a higher plane. This alternation of regression and progress until the final achievement of the *summum bonum* suggests an evolution of the soul, and presumably dictates the content and the sequence. of the sonnets.

The sonnets may be described as falling generally into five categories, but the distinction is not sharp. Each category is in some degree present in every other, but each has its distinct emphasis within the frame of the whole. The sequence, then, develops organically, in accordance with the psychic evolution of which it is the vehicle.

In the first group of sonnets we may place those which describe the lover's initial struggle, the conflict of flesh and spirit after the first vision of beauty.[1] The lover has seen the ideal beauty, but, as Bruno tells us, has been scorned as unworthy of it.[2]

All the sonnets of De gli eroici furori are written in the inflated hyperbolic style of the later Petrarchists, with great exuberance of imagery and conceit. As much may be said of the sonnets of Bruno's contemporaries, Luigi Tansillo and Antonio Epicuro, and, generally speaking, of the other poets of the Neapolitan school which influenced him. However, since Bruno's sonnets are full of esoteric symbolism and consciously directed to a learned and highly exclusive audience, the expression is frequently cryptic in the extreme. The sonnet "Io che porto", for example, begins:

> I who carry the lofty banner of love, have frozen
> hopes and burning desires: at one and the same time
> I tremble, freeze, burn and sparkle, I am dumb and I
> fill the sky with ardent shrieks.
>
> My heart throws off sparks, while my eyes
> distil water; and I live and die, laugh and
> lament...[3]

Like the lover of Petrarch's sonnets, Bruno's lover burns and freezes, is struck dumb and wanders about restlessly, filling the world with lamentations. The intensity of his animal passions causes him to burn in anguish while the intellectual passion chills him and freezes him into inactivity. He hopes continually for spiritual deliverance, and is plunged into despair because of the persistently violent demands of the flesh.[4]

The sonnet "Bene far voglio" symbolizes the beloved as the radiant sun, whose celestial light inspires the lover and yet eludes him because of his ignoble instincts:

---

[1] GIORDANO BRUNO, Opere italiane, 352-356; my translation, pp. 99-104; see also MARSILIO FICINO, Opera omnia (Basel, 1561), pp. 657-658.

[2] Ibid., and my translation, p. 101. For Bruno's suggestion in the closing dialogue that the lover's rejection by the lady, Giulia, was eventually the means by which he was raised to divine contemplation, see below, p. 258 and note 1, and pp. 266-267.

[3] BRUNO, Opere, 1, 2, 348; my translation, p. 96.

[4] Ibid., 1, 1, 353-354; my translation, p. 99-100.

I long to do good, but it is denied me; my sun is
not with me, although I am with it; for in order to
be with it, I am no longer with myself, and the nearer I am
to it, the further it is from me.

For one moment of joy, I do much weeping; seeking
happiness, I find affliction; because I look too high,
I am blinded; and to obtain my good, I lose myself. [5]

The sonnets of the first group describe the lover's torment. In
the second group the lover has become aware of the true nature
of his lady and of the means necessary to possess her. Reminiscent
of the lover in the poetry of the *stil novisti* he understands now
that the beauty which defies him has attributes of a divine nature
which he can embrace only by purging his soul of its baseness.
The lover's new insight is expressed in the sonnet. "Venere, dea
del terzo ciel":

Venus, goddess of the third sphere, and mother
of the blind archer, subduer of all men; that
other, sprung from the forehead of Jove, proud
wife of Jove, Juno,

call the Trojan shepherd, to judge which of them,
most beautiful, deserves the golden fruit. If my
goddess were set among them, it would be awarded
neither to Venus, Athena or Juno.

The Cyprian goddess is beautiful by reason of
her lovely limbs, Minerva through her intellect,
and Juno pleases by that worthy splendor

of majesty, which satisfies the Thunderer; but
my goddess contains within herself all that is requisite
of beauty, intelligence and majesty. [6]

Thus he realizes, so we are told in the commentary, [7] that his
ideal lady transcends the limited and purely finite virtues of

---

[5] *Ibid.*, 2, 1, 449; my translation, p. 201.
[6] *Ibid.*, 1, 5, 417-418; my translation, p. 167.
[7] *Ibid.*, pp. 417-421; my translation, pp. 168-170.

beauty, intelligence and majesty, for she includes all three beauties as modes of one eternal being. Now the lover begins to see the lady as an enigmatic substance which contains irreconcilable contraries: the individual and the universal, the many and the one. The beloved, therefore, begins to reveal to him characteristics which the lover may also apply to the highest of "monads", which, although finite in appearance, contains the divinity. [8]

"Chiama per suon di tromba il capitano" tells of the struggle of the lover to organize the powers of his soul in order to attain the beloved fully:

> The captain with the sound of the trumpet
> summons all his warriors beneath one banner; where,
> if it happens that for some of them it sounds in vain,
> and they come not promptly,
>
> those who are traitors he kills, the mad men he
> banishes from his camp or he scorns them: so the soul
> with those of its intentions which come not to assemble
> under one standard, either it wishes them dead or removed.
>
> I regard one object, which absorbs my mind,
> and it is a single visage. I remain fixed
> upon one beauty,
>
> which has so pierced my heart, and is a single
> dart; for by one flame only I burn, and know
> but a single paradise. [9]

The symbolism here is transparent; nevertheless, Bruno explains it fully. The captain is the lover's will governing the appetites and affections through reason. The enemies he purges are those violent animal passions that keep him from the beloved ideal, although he holds her in view. [10] Here Bruno follows Dante

---

[8] *Ibid.*, 2, 2, 473; my translation, pp. 225-226. For a discussion of the monad as an entity which though finite in appearance contains the infinite, see GIOVANNI GENTILE, *Giordano Bruno e il pensiero del Rinascimento* (Firenze, 1920), pp. 79-85. See also BRUNO, *De la causa*, in *Opere* 2nd ed. Gentile, I, 248-260.

[9] BRUNO, *Opere*, II, 1. 1. 339; my translation, p. 86.

[10] *Ibid.*, p. 340; my translation, pp. 87-88.

and makes the beloved the spur to a higher stage of spiritual progress.

In another group of sonnets we are shown the lover at the other extreme, in danger of becoming totally absorbed by his ideal, even at the expense of the valid demands of his corporeal nature. [11] He would remain, then, wholly in the ascetic half of his soul's circle and neglect the corporeal half to his detriment:

> The youthful Actaeon unleashes the mastiffs and the greyhounds to the forests...
>
> Here among the waters he sees the most beautiful countenance... that ever mortal or divine may see... and the hunter becomes the prey that is hunted...
>
> I stretch my thoughts to the sublime prey, and these spring back upon me, bring me death by their hard and cruel gnawing. [12]

According to the commentary, the youth Actaeon represents the intellect in pursuit of divine knowledge. The greyhounds and the mastiffs represent the will and the appetites that are unleashed in the hunt for that divine goodness and beauty which the intellect is more likely to love than to comprehend. [13] But so ascetic is the desire of the intellect that it fails to take into account the valid demands of physical nature without which the individual cannot survive. [14] Thus, the lover's soul, intent upon intellectual contemplation, experiences a violent cleavage within him, and his desires gnaw him cruelly and severely. [15]

---

[11] *Ibid.*, 1, 3, 360; my translation, pp. 108-109.
[12] *Ibid.*, 1, 4, 374, 381; my translation, p. 123.
[13] *Ibid.*, 1, 3, 360-361; 1, 4, 375-385; my translation, pp. 109, 123-127.
[14] *Ibid.*, 1, 4, 381, 386; my translation, pp. 129-137.
[15] WILLIAM SHAKESPEARE makes Orsino describe himself in analogous terms in *Twelfth Night,* I, i, 19-24.

> O when mine eyes did see Olivia first,
> Methought she purged the air of pestilence.
> That instant was I turned into a hart,
> And my desires, like fell and cruel hounds,
> E'er since pursue me...

In the sonnet, "Alti, profondi e desti miei pensieri", a complex of audaciously unconventional metaphors, the lover understands that he must balance his asceticism by the legitimate demands of his corporeal nature. The poem is conceived in the mold of the *cor-anima-mente* sonnets of the *stilnovisti,* and involves mainly the conflict of the eyes and the heart:

> Lofty, profound and living thoughts of mine,
> which desire to flee the maternal bonds of the afflicted
> soul, and are disposed as archers to aim at the
> point where the lofty idea is born.

> Arm yourselves with the love of the domestic
> fires, and curb your eyes so forcefully, that

> these companions of my heart, should not make
> you strangers to it. [16]

How, Bruno asks, can the soul be at home in the body, beside its "domestic fires", when his thoughts abandon it in the pursuit of immaterial and divine nourishment only? If the thoughts neglect the body to this extent, the body will starve. Here Bruno departs considerably from the metaphysics of the poetry of the *stil novisti,* for he tells us that the domestic fires of the body are the natural instincts of human love and generation without which it is impossible to become a fully integrated soul. [17]

The neglect of the corporeal for the exclusively spiritual is again taken up at the end of the book in the lament of the blind men who suffer because they have been too preoccupied by the divine light. The fourth blind man says:

> ...Thus I remain with my spirit all intent
> upon the most living light which illumines the
> world, and I am insensible toward all the lesser

---

[16] Bruno, *Opere,* I, 4, 382; my translation, pp. 130-131. For further discussion of the manner in which the *stilnovisti* employ eyes, heart and mind as allegorical antagonists in a "new" drama of the soul, see Maurice Valency, *In Praise of Love* (New York; Macmillan, 1958), pp. 230-233.

[17] *Ibid.,* pp. 385-387; my translation, pp. 134-136; see also above, p. 25 and notes 9 and 10.

splendors: and while this light shines upon the
world, it willingly pays attention to no others. [18]

Although he is now aware that he must balance the demands
of the intellect with those of his physical nature, the impassioned
lover, in the next group of sonnets, is nevertheless unable to
avoid becoming totally absorbed by the ideal. He would scorn
the life of the body to find new birth in the life of the spirit. [19]

The sonnet "Se la farfalla" describes the state of mind of the
lover who would willingly consume his flesh in order to become
absorbed by the one eternal fire:

> If the butterfly wings its way to the sweet light
> that attracts it, it is because it knows not that the
> fire is capable of consuming it; if the thirsty stag
> runs to the brook, it is because he is not aware of the
> cruel bow...
>
> If my languishing is so sweet to me, it is because
> of the heavenly face that delights me so, and because the
> heavenly bow so sweetly wounds;
>
> and because in that knot is bound up my desire, I
> suffer eternally through the fire in my heart, the arrow
> in my breast, and the yoke upon my soul. [20]

These are perfectly commonplace comparisons. The lover is
compared to the butterfly drawn to the fire and to the stag
wounded by the hunter's bow. [21] As Bruno tells us in the commentary, because the butterfly and the stag are not aware of the
ruin to be inflicted upon them, they follow blindly the instincts
of their corporeal natures. But the lover knows he must suffer

---

[18] *Ibid.*, 2, 4, 492-494; my translation, pp. 245-246. For William
Shakespeare's analogous statement see Berowne's speech, *Love's Labour's
Lost*, I. i, 72-79.

[19] *Ibid.*, 2, 1, 447-448; my translation, pp. 198-200.

[20] *Ibid.*, 1, 3, 360-363; my translation, pp. 110-112.

[21] For Pierre de Ronsard's allusion to the lover as a stag wounded by
the hunter's bow see son., "Comme un chevreuil". For William Shakespeare's similar allusion see *Love's Labour's Lost* IV, i, 35-58 and IV, ii,
1-67.

physical pain in order to attain eternal life. Therefore, he welcomes the knot of love for it symbolizes the mastery that he wishes his soul to exert over his body. [22]

The desire to annihilate the body in order that the soul may rise again upon its ashes is emphasized in the sonnet "D' un si bel fuoco":

> By so beautiful a fire and so noble a yoke,
> beauty enkindles me, and chastity entangles me, so
> that I must be happy in fire and in slavery...
>
> Because so lovely a flame enkindles my heart,
> and the desire for so sweet a bond compels me,
> darkness is my servant and my ashes glow. [23]

In the sonnet "Per man d'amor" the lover addresses the divine lady, and in the usual paradox he asks for the death that is also the key to life:

> If you condescend that I may live, open the
> gates to your gracious glance; gaze upon me, oh
> lovely one, if you wish to give me death. [24]

In the next group of sonnets the lover finally abandons the ideal of complete denial of his corporeal nature in order to achieve unity with the Deity. Bruno returns to an idea which becomes his final philosophy of the soul. The lover sees that such a denial creates an artificial cleavage between the corporeal and the spiritual, a dichotomy that does not in fact exist in the true nature of the soul, for the corporeal and the spiritual elements complement each other to form the whole. [25] The lover now understands that to achieve the supreme good he must not deny

---

[22] *Opere*, II, pp. 360, 368; my translation, pp. 108-111.

[23] *Ibid.*, 1, 3, 363-365; my translation, p. 111.

[24] *Ibid.*, 2, 1, 447; my translation, p. 199.

[25] *Ibid.*, 1, 4, 386-387; my translation, pp. 134-135. For the statement that in the individual BRUNO conceived a material principle of equal value with the formal principle see ANTONIO CORSANO, *Il pensiero di Giordano Bruno nel suo svolgimento storico*, ed. G. C. Sansoni (Firenze, 1940), pp. 216-221.

the body, but rather he must bring it into harmony with the spirit. This he may achieve by means of the rational faculty that synthesizes the opposites and lies at the center of the soul. [26]

The sonnet "Annosa quercia" expresses the lover's satisfaction in this ability to unite the forces of his corporeal and spiritual natures toward the desired end:

> Ancient oak, which spreads its branches to the air, and fixes its roots in the earth; neither the trembling earth, nor the powerful spirits which the sky lets loose from the bitter north wind... can ever uproot you from the place where you stand firm...

> ...You demonstrate the true portrait of my faith, which no external accident has ever shaken...

> Upon one single object I have fixed my spirit, my sense and my intellect. [27]

This conceit, according to the commentary, describes the constancy that the lover achieves despite the tempestuous onslaughts of the lower passions. [28] He holds fast his sense, his intellect, and his spirit toward the beloved object. In this state of tranquillity he has desire without any sense of pain, the highest beatitude in this mortal life. [29]

The lover has now found a new harmony within himself. He also discovers that an analogous harmony may be found in the external universe. He realizes that in nature the harmonious relation between its finite and infinite aspects fully satisfies his intellectual yearning. In this harmony he finds the *summum bonum*. [30]

In the next group of sonnets the lover sees that he is vitally related to the outer universe because it contains, as does his own nature, a twofold substance: a synthesis of the corporeal and the spiritual, the material and the formal, the transient and the perma-

---

[26] *Opere*, II, 1, 3, 372; my translation, p. 121.
[27] *Ibid.*, 2, 1, 5, 411-412; my translation, p. 162.
[28] *Ibid.*, pp. 413-415; my translation, pp. 162-164.
[29] *Ibid.*, p. 413; my translation, p. 163.
[30] *Ibid.*, 2, 2, 437; my translation, pp. 225-226.

nent, both of them aspects of the one eternal being.[31] This dual substance he figures in the symbol of Diana. She is Bruno's *philosophia*, as the *Donna Sapienza* was Philosophy for Dante in the tractates of the *Convivio*.

The sonnet, "Venere, dea del terzo ciel" included in the category discussed above deserves a place in the present group since it foreshadows the lover's final understanding and attainment of the *summum bonum*. This is the sonnet in which the lover first becomes aware of the paradoxical nature of his beloved ideal as containing three modes of the one eternal being.[32] Now in the final stage of his progress he knows this goddess no longer as Venus, but as Diana, the Monad, the highest finite mode of the infinite being. As he had begun to perceive earlier, she alone is the all-beauty, truth and power, and these three are but finite modes of the One.

The sonnet "Forte a' colpi d'Amor" alludes to Diana as the sacred light which has found easy entrance to his heart:

> Strongly I waxed in virtue under the blows
> of love...
>
> At last, one day (as the heavens destined it)
> I found myself so fixed by those sacred glances,
> which through my eyes and alone among all others,
> found easy entrance to my heart.[33]

The sacred rays that emanate from the lady's eyes are here equated with the Plotinian superessential entity, which is the source of ideas. However, the source of ideas cannot be known absolutely.[34] Accordingly, Bruno, in the commentary, personifies

---

[31] See BRUNO, *De la causa*, in *Opere*, I, 177-196, for Bruno's view that the material and formal principles, though distinct, are in reality but two manifestations of the one infinite being. For further discussion of Bruno's view of reality as a unity of the material and formal principles in the infinite, see CORSANO, *Il pensiero di Giordano Bruno*, pp. 220-221, and JULIE SARAUW, *Der Einflus Plotins auf Giordano Brunos De gli eroici furori* (Leipzig, 1916), p. 55.

[32] See above, pp. 34-35.

[33] BRUNO, *Opere*, II, 2, 1, 453; my translation, p. 206.

[34] Bruno follows Plotinus and holds a negative theology. Nicolaus Cu-

the source of ideas as Apollo, the primary intelligence, and its manifestations in the intelligible world as Diana, the secondary intelligence, which reflects the splendor of the first. [35]

The lover arrives at full contemplation of Diana in the sonnet, "Chi femmi ad altro amor"; here he enjoys the full ecstacy of his vision:

> She who kindled my mind to the higher love,
> she who rendered every other goddess base and vain
> to me; she in whom beauty and sovereign goodness
> are uniquely displayed,
>
> is she whom I saw coming from the forest,
> huntress of me, my Diana, among the lovely nymphs
> upon the golden Campania, wherefore I said to love:
> —I surrender myself to this one.
>
> Captive though I am in her amorous court,
> I am so highly blessed, that I do not envy
> the freedom of any man or god. [36]

In the commentary Bruno tells us that Diana is chief among the lesser goddesses because in her resides the multitude of species, forms, and ideas. [37] She is the truth, beauty, and goodness which men see only partially as they pursue their various modes of life. Diana is the Plotinian νοῦς realizing itself in the world of phenomena. [38]

The lover knows that the partial truths are modes of the one that pervades all. He is no longer the captive of merely transient

---

sanus and Marsilio Ficino held a negative theology similar to that of Plotinus. Accordingly, the Deity in his own proper essence is unknowable. See PLOTINUS, *Enneads*, 5. 4. 1., and NICOLAUS CUSANUS, *Vision of God*, trans. G. E. Salter (London, 1928), pp. 58-59. See also PAUL O. KRISTELLER, *The Philosophy of Marsilio Ficino*, trans. Virginia Conant (New York, Columbia University Press, 1943), pp. 240-245.

[35] *Opere*, 2, 1, 452-453; my translation, pp. 204-205. See also PLOTINUS *Enneads* 5. 7.

[36] *Opere*, II, 2, 2, 463; my translation, pp. 216-217.

[37] *Ibid.*, p. 464; my translation, p. 217.

[38] PLOTINUS 5. 7. In his concept of the world of phenomena, however, Bruno differs considerably from Plotinus. Bruno holds matter is metaphysically subsistent. In the view of Plotinus matter is devoid of quality and is unilluminated privation; see PLOTINUS *Enneads* 2. 4. 3, 6, 7, 8, 9.

allurements, nor of sporadic asceticism, which is also partial. Both the spiritual and the corporeal elements are necessary for the fullness he seeks, since they are both necessary to complete the circle of the external universe of which he is a part. As the truth pervading all the modes in the world of phenomena, Diana is the limitless center that holds them together.

The lover sees Diana, then, as the highest Monad, in whom is united the contraries of finite and infinite, of the many and of the one:

> ...And now he sees everything as one, not any longer
> through distinctions and numbers, as according to the
> diversity of the senses, or as varied fissures [39] are seen
> and apprehended in confusion. He sees the
> Amphitrite, the source of all numbers, of all
> species, which is the *monad*, [40] the true essence
> of the being of all things; and if he does not see it
> in its own essence and absolute light, he sees it in
> its germination which is similar to it, which is its
> image: because from the monad which is the divinity,
> proceeds this monad which is nature, the universe,
> the world; where it is comtemplated and gazed upon as
> the sun is through the moon, which is illuminated
> by it, inasmuch as he finds himself in the hemisphere
> of the intellectual substances. She is Diana, that
> one which is the same being, that being which is
> the same truth, that truth which is intelligible nature,
> in which is infused the sun and the splendor of the
> superior nature, according as the unity is distinct
> in that which is generated, and that which generates,
> or that which produces and that which is produced. [41]

The lover, then, comprehends that Diana is the source of unity of the material and informing principles in the infinite, and that, finally, she is the universe as a finite mode of the Deity.

The lover has now reached the full realization of the ideal. But he has attained it only in proportion to the harmony he has brought about within himself. For only now is he able to realize

---

[39] *Opere*, II, 2, 473 and note 2.
[40] Italics mine. See above, p. 35 and note 8.
[41] *Opere*, II, 2, 2; my translation, pp. 226-227.

an adequate relation between his own center and the center of the universe.

The last poems in this category are those which bring the last dialogue to a close. These are no longer sonnets but *canzoni*, whose form according to the *capceudadas* rhymes brings the whole work to full circle. [42]

The "Canzone de gl' illuminati" is sung in honor of Diana by nine lovers as they dance in a circle. The lovers celebrate Jove, ruler of the firmament, as the superessential intelligence, and Father Ocean as ruler of the empire of nature. [43] In their song, Jove and Ocean dispute as to which of them is the supreme power, and the dispute is resolved when Jove declares that, while no power is superior to his, yet Ocean, which numbers among his nymphs one who is as splendid among the stars as is the sun in the sky, is of equal power with himself. This resplendent nymph is Diana, who is enthroned upon the Thames, and she, of course, represents Queen Elizabeth to whom the first sonnet of *De gli eroici furori* is dedicated. Thus the philosophical allegory is turned, as in the *Faeree Queen*, into a compliment to the most splendid of living ladies, the goal and ideal of all human aspiration:

> And Jove replies: —Oh god of the tossing seas,
> that anyone be found more blessed than I is not
> allowed by fate; but my treasures and yours run
> their course together.

> The sun avails among your nymphs through this one;
> and by the force of eternal laws, and of the alternate
> abodes, she is valued as the sun among my stars. [44]

It is evident that Bruno adapts from the traditional sonnet sequence those elements which are most convenient to his philo-

---

[42] See below, pp. 73-74, 76-77, 263-265.

[43] *Opere*, II, 2, 5, 517 and note 2.

[44] *Ibid.*, pp. 518-519; my translation, pp. 265-266. For similar allusions to Queen Elizabeth of England, see "Argomento del Nolano", pp. 316-317, in my translation, p. 65 and note 8; and "Iscusazion del Nolano", p. 330; my translation, "Apology of the Nolan", p. 79 and note 2.

sophical purposes: the play of antithesis characteristic of the Petrarchan sonnets as interpreted in the prose commentaries upon them and in the Platonic *trattati d'amore,* the theme of the lover's progress as he frees himself from the conditions of carnal love in order to attain the object of spiritual desire, the use of the lady as a symbol for the ultimate object of aspiration. But Bruno's Diana has, of course, primarily a metaphysical connotation. She is a synthesis of the Plotinian νοῦς and the divine principle of intelligibility existent in nature.

The Petrarchan dialectic of contraries expressing the warring passions of the lover as he burns and freezes, fears and hopes, is especially useful to Bruno. It is made to reflect his concept of the soul as an entity in which the material and the formal principles struggle for mastery. Bruno brings, as a matter of fact, the Petrarchan dialectic to its metaphysical end. For Bruno the solution lies in a renunciation of neither the corporeal nor the spiritual element, but in bringing both into harmony as his conception of the soul requires. Thus he chooses a middle path toward the ideal. The lover must renounce excessive preoccupation with either the body or the soul in order to contemplate Diana, the finite mode of the infinite being.

In Dante and Petrarch the lover may overcome his internal conflict by transcending the conditions of corporeal love, so that his soul may be set free for union with a knowable God. As in Dante and Petrarch, and in the *trattati d'amore,* Bruno's lover at first suffers privation of his beloved object, but he finds it again, after a long, laborious process, in its ideal form. However, the death of the beloved is not necessary to aid the lover's progress toward the *summum bonum.*

In Bruno's view, the beloved object is not of the same lineage as Beatrice and Laura. Diana is closer to the Platonic idea and the Plotinian νοῦς than to the angelic nature of Beatrice. Accordingly, Diana does not need to die in the body and ascend to heaven in order to inspire her lover from beyond the *primum mobile.* Bruno's cosmology admits no order of spheres ascending from earth to the Empyrean heaven. For Bruno the universe contains innumerable planets, including the earth, unlimited in space and time. [45] Diana is its finite and eventually its infinite manifestation.

The ideal lady, then, in Bruno's *De gli eroici furori* is no less related to the cosmos than the lady in the poetry of the *stil novisti*. Beatrice and Laura, however, belong undeniably to a universe essentially geocentric, while Diana belongs to a universe that is essentially heliocentric as well as infinite, as Bruno's cosmology demands.

---

[45] BRUNO, *De l'infinito universo e mondi*, in *Opere*, I, 340, 401-402.

IV

DE GLI EROICI FURORI AND THE EMBLEMATIC TRADITION

The emblems of *De gli eroici furori,* following the example of the sonnets, allegorize the lover's arduous search for the ideal. As the sonnets reflect Bruno's philosophy of love, so the emblems reflect it also in their way.

In the first four dialogues of *De gli eroici furori* Bruno's spokesman, Tansillo, bases his commentary solely upon the sonnets which he quotes. Early in the fifth dialogue, he adopts a new technique; he grounds his commentary on an album of emblems in which are expressed the various states of the lover as he struggles toward the *summum bonum.* Bruno gives special importance to the emblems in the total design of *De gli eroici furori.* Emblems introduce the sonnets exclusively in the fifth and last dialogue of the first part and in the first of five dialogues in part two. Bruno, accordingly, designs them to occupy the precise center of *The Heroic Frenzies.* Their symbolism, therefore, is central to the work as a whole. These emblems, however, are described, not pictured; and each has a Latin motto which is explained by a sonnet and further elucidated by the commentary.

The habit of elucidating a series of emblems by a sonnet or commentary had been in vogue for almost one hundred years before Bruno published *De gli eroici furori.* In the sixteenth century the emblem book was an immensely popular form of literature in Europe, and many were published. These books contain a series of black and white drawings, each of which has its own motto in either Latin or French, and each is usually followed by a poem that develops the thought behind the emblem. The series as a whole illustrates a theme of a personal, political, ethical or theological nature, so that an emblem book has a certain unity.

The emblem books of Paolo Giovio, Girolamo Ruscelli, Andrea Alciati and Theodore de Bèze were especially popular during the period preceding the publication of *De gli eroici furori*. They have, therefore, a special relevance to the peculiar way Bruno employs his emblems.

Giovio and Ruscelli alone, fortunately, provide some *rationale* which allows us to conclude that on the whole emblem writers of the sixteenth century quite consciously intended their emblems to have other than literal meanings. In the second half of Paolo Giovio's *Ragionamento sopra impresa* there is a valuable critique, which analyzes the origin of emblems, their nature and the levels of meaning that can be attached to them. Neither Giovio nor Ruscelli mention these levels specifically. But the meanings they discuss are noticeably analogous to the traditional system of allegory in the Renaissance, the four levels, literal, allegorical, tropological and anagogical. [1]

In the *Ragionamento* Giovio suggests that figures of leaves, trees, smoke, water, mechanical devices, and even grotesque animals and birds be employed in such combinations as will make the emblem pleasant and beautiful. The properly designed emblem will express a logical relation between the concrete figure and the meaning conveyed. [2]

In a good emblem the symbol, the idea, and the motto must combine in a harmonious manner. Giovio's demanding aesthetic sense recognizes that the perfect emblem is rarely achieved, and he gives several examples of emblems and mottoes which in his view are for one reason or another imperfect. [3]

Giovio points out further that the ideal emblem must neither be totally obscure nor too obviously intelligible. If it is totally obscure, no one, ignorant or wise, will understand it;

---

[1] DANTE *Convivio* 1. 1; see also PIERRE DE RONSARD, *L'Abrégé de l'art poétique francois*, ed. Blanchemain, in *Oeuvres complètes* (Paris, 1866), VII, 317-319.
[2] PAOLO GIOVIO, *Ragionamento sopra imprese* (Venezia, 1566), pp. 6-7.
[3] Giovio, pp. 10-12.

if too obviously intelligible, it will offend the wise. The perfect emblem, therefore will steer a course midway between the two extremes. It will be intelligible to all but the ignorant, yet it will be sufficiently esoteric to stimulate the wise to seek out its inner meaning. To this end, the motto and the verses, he suggests, "ought to be written in a language different from the author's own customary idiom". [4]

Giovio's insistence upon a degree of obscurity suitable to the emblem is another example of a common Medieval and Renaissance attitude with respect to poetic symbolism in general. Dante, Petrarch, Boccaccio, as well as others, hold that some degree of obscurity in the symbol is necessary to veil the *sovrasenso*. Petrarch tells us the poet's function is to "veil the fundamental truth within a delightful, many-colored cloud, preparing the reader for a long but sweet labor". The more difficult it is to search for the meaning, the sweeter it is when found. Quite clearly Petrarch has in mind the pleasure of obscurity in the symbol analogous to the pleasure Saint Augustine finds in the figurative obscurity of Scripture. The suggestion seems to be that the wise are elected to penetrate the most profound truths primarily by the silent communication of the *sovrasenso*. [5]

In Part II of the *Ragionamento*, Giralomo Ruscelli develops several levels of meaning which the emblem may involve. [6] The Greek word, ἔμβλημα, he says, at first referred simply to devices found on the shields of Greek warriors or on coins and mixing bowls. They were designed for the purpose of ornament as well as for the expression of some national or political end or ideal. The Roman word, *emblema*, had a similar meaning, but was usually associated with some momentous war-like undertaking.

---

[4] *Ibid.*, pp. 6, 7.

[5] St. Augustine *De doctrina christiana* II, 6, 7; Petrarch, *L'Africa*, ed. Ricciordi, ix, in *Letteratura italiana* (Milan, Naples, n. d.), VI, 695-696, 11. 65-101; Boccaccio, *Genealogie deorum gentilium libri*, ed. V. Romano xiv-xv, in *Scrittori d'Italia* (Bari, 1951), II, 699-700, 767, 768; see also Anthony J. Mazzeo, "St. Augustine and the Rhetoric of Silence", *Journal of the History of Ideas*, XXIII: 2 (1962), 175-196.

[6] The *Discorso di Girolamo Ruscelli intorno all'inventioni dell'Imprese, dell'Insegne de' Motti, et delle Livree* occupies pages 113-236 of Giovio's volume.

Actually, he says, the Roman word, *expeditio*, was more commonly used in this connection, and especially with reference to military undertakings. So, in the *Aeneid*, Virgil uses the word *expeditio* in connection with the devices worn on the Greek warriors in the campaign against Troy. According to Ruscelli, the Italians of the fifteenth and sixteenth centuries adapted the Roman *expeditio* to their own purposes. The Italian word, *impresa*, thus came to mean not only "enterprise" or "undertaking", but also the device or emblem associated with such an undertaking. *Impresa* was derived from the Italian verb, *imprendere*, "to undertake or to desire to do a thing with the obstinate intention of seeing it through to its accomplishment". [7] The device, therefore, symbolized a lofty determination of historical or political character, and the emblems used by the various kings and dukes of Italian cities were generally of this nature.

The *impresa*, moreover, would ordinarily involve several levels of allegory, beyond the obvious historical implication. The historical was considered the literal meaning, but, as Ruscelli suggests, the *impresa* could express higher degrees of meaning also. "The intention to see an undertaking through to its accomplishment", he tells us, may refer not only to a warrior's will to sacrifice himself for the glory of his nation, but also to the lover's will to sacrifice himself in the pursuit of the beloved object. Thus the *impresa* is especially useful to "those noble, valorous and gentle spirits" who pursue the "lofty end of love and glory". [8] Ruscelli refers to Petrarch specifically as a poet who employs the word, *impresa*, in relation to the lover's secret will:

...Imprese, then, refers to all grand things of
great moment with which Princes and Magistrates occupy
themselves, as with war. Thus we speak of the great
impresa against the Turks.

...Impresa, of course, may have a *private* meaning, [9]
as it had in the sense in which Petrarch was wont
to use the word... [10]

---

[7] Giovio, p. 178.
[8] *Ibid.*, pp. 181-182.
[9] Italics mine.
[10] GIOVIO, pp. 178-179.

He then quotes a passage from Petrarch the first two lines of which appear in the seventy-fifth sonnet:

> Questi son quei begli occhi ch'Imprese
> Del mio Signor vittoriose fanno...
> Tanti ti prego, piú gentile spirto,
>
> Non lasciare la magnanima Impresa,
> Piaciati omai col tuo lume, ch'io torni
> Ad altra vita e à piú belle Imprese... [11]

The passage from Petrarch plays upon the word, *impresa,* so that it suggests a sense beyond the literal. The lady's radiant eyes are a means by which God's purpose becomes triumphant over the lover. In the eyes of his lady, therefore, the lover reads the *impresa* of God, whose secret will is eventually to enlist him under His ensign. Ruscelli suggests, then, that Petrarch uses the word *impresa* to describe essentially the same purpose intuited by the lover in the poetry of the *dolce stil nuovo.* Dante in the *Vita nuova* gave the honored lady of the courtly tradition the metaphysical role of intermediary between her lover and God. Moreover, the lover intuited that the divine ray reflected in the eyes of Beatrice was the means by which her role was to be fulfilled. [12]

Ruscelli's interpretation of the emblem throws interesting light on the various levels of meaning Paolo Giovio, Andrea Alciati, Théodore de Bèze and Giordano Bruno have given their emblems.

The *Ragionamento* of Paolo Giovio was published in Venice in 1566 and was translated into English by the Elizabethan poet,

---

[11] *Ibid.,* p. 179:

> These are the eyes which make the Enterprises of my
> Lord triumphant...
> So much do I beg you, most gentle spirit, do not
> abandon your magnanimous Enterprise,
> May it please you now by your light, that I turn to
> that other life and most beautiful Enterprises...

The last four lines are variants from Ruscelli's text and do not correspond to the critical edition of Petrarch.

[12] DANTE *Vita nuova* 14. 20.

Samuel Daniel, in 1584 while Bruno was in London. [13] This book of emblems is not a traditional one of drawings, mottoes and sonnets; it is primarily an analysis of the *imprese* and mottoes employed on the seals of various dukes and princes of Italian and other European city-states. The emblems are discussed without being illustrated, since Giovio was concerned chiefly with the effectiveness with which they expressed political or military objectives.

The *Emblematum* of the Italian jurist, Andrea Alciati, were widely published from 1534 to 1616 in Milan, Venice, Paris and especially by the Bonhomme press in Lyon. [14] The *Emblematum* contains a series of emblems divided into three conventional categories. One series represents the vices, another the virtues, and the third represents miscellaneous subjects. Most of the emblems are followed by a Latin motto and a gloss in verse.

The emblem book of Théodore de Bèze, *Les Vrais Portraits des hommes,* published in 1581, is Calvinistic in its orientation. [15] His emblems glorify a number of clerics who were stigmatized as heretics by the Roman Church. Each emblem is followed by a short verse in French, but is not accompanied by a motto. De Bèze's book has a refreshing clarity and a succinct quality that, unhappily, is absent from most emblem books of the period, the

---

[13] DOROTHEA SINGER, *Bruno His Life and Thought* (New York, 1950), pp. 37-38.

[14] The following editions apparently enjoyed immense popularity:

> 1531, *Emblematum,* published by Sentry Steyner, Augsburg Press.
> 1534, by Christian Wechsel, Paris.
> 1536, by Wechsel in Paris, the first French translation.
> 1542, by Wechsel in Paris, the first Latin translation.
> 1544, by Jacques Moderne at Lyon.
> 1546, by the Aldus Press in Venice.
> 1548-1616, by Nathias Bonhomme at Lyon in Latin, French, Spanish and Italian, with woodcuts by Le Petit Bernard.
> 1566, by George Roben and Sigismund Beyerbabend at Frankfort.
> 1583, Nicholas Basse, Frankfort, Germany.
> 1549, an Italian version of an earlier edition owned by Edward VI of England.

[15] Théodore de Bèze joined Calvin in Geneva in 1548 and became his adjutant in 1549. The edition, *Les Vrais Portraits des hommes,* bears no place of publication on the title page.

authors of which made a virtue of obscurity and esoteric mysticism.

Of the twenty-eight emblems in De gli eroici furori sixteen compare rather closely with the emblems of Bruno's contemporaries. Twelve of the sixteen appear to be modifications of emblems employed by Andrea Alciati, Théodore de Bèze, and Paolo Giovio. Four reproduce closely specific emblems of the same contemporaries.

The remaining twelve emblems that Bruno uses seem to have little or no relationship to those of his contemporaries, and it is possible that he may have devised them himself. Or he may have chosen them from emblem books that are no longer available to us. Beyond a doubt, Bruno was familiar with a number of alchemical works circulating freely through Europe during the sixteenth century. In these works alchemical symbols are widely used to interpret the mysteries of revealed religion. Several of these suggest emblems that we find in De gli eroici furori. A striking example is Bruno's adaptation of the mandala symbol [16] found in alchemical manuscripts. [17] Varied forms of the mandala are seen in Bruno's emblem of the burning arrow over which is a circular noose, in his emblem of the flaming yoke circumscribed by a noose, and especially in his emblem of the wheel of fixed time moving about its own center.

For the sake of convenience, Bruno's emblems may be considered in three groups. The first group consists of those emblems in which Bruno follows Paolo Giovio; the second comprises those emblems in which he follows Andrea Alciati; in the third group are those in which he follows Théodore de Bèze. [18]

---

[16] The Sanskrit word, "mandala", means a magic circle, and its symbolism includes concentrically arranged figures. "It is one of the oldest religious symbols (the earliest form being the sun-wheel), and is found throughout the world." FRIEDA FORDHAM, An Introduction to Jung's Psychology (Penguin, 1953), 65. See also below, p. 264 for Bruno's adaptation of the mandala symbol both for the theme and structure of the final canzoni.

[17] For the frequent use of the mandala symbol by the alchemists, see HERBRANDT JAMSTHALER, Viatorum spagyricum (Frankfort, 1625), p. 272; see also THOMAS AQUINAS, pseud., De alchimia (Codex Vossianus, Leyden, 16th century), p. 74.

[18] For a detailed comparative analysis of Bruno's emblems with those of Paolo Giovio, Andrea Alciati and Théodore de Bèze, see my article,

A careful study of the three groups makes clear beyond a doubt the peculiarly esoteric use to which Bruno put the emblems he derived or adapted from others. Giovio's emblems express a meaning that corresponds primarily to the literal-historical. The emblem of Henry II of France, a full moon, symbolizes the auspicious beginning of his political reign which will give light and glory to the world. [19] In Andrea Aliciati's emblems may be found a meaning mainly corresponding to the tropological level, but in conspicuously conventional sentences. The emblem of a ship at sea with two men leaning lackadaisically over the rail indicates, as the motto tells us, simply that they "withdraw easily from virtue". [20] As Théodore de Bèze's emblems were religious in intention, the meaning in general suggests an anagogical implication. Accordingly, his emblem of a phoenix rising upon its ashes represents, as the verse explains, the Calvinistic saint who will rise again after his martyrdom by heretical hangmen, and it suggests the ultimate triumph of true religion. [21] For Giordano Bruno, however, the emblem symbolizes very fancifully a phase of the relentless struggle of the aspiring lover to achieve the infinite good, even while that good remains persistently elusive.

Several of Bruno's emblems which elaborate upon the lover's arduous search for an evasive ideal are particularly relevant. Bruno's emblem of a full moon represents the heroic lover's potential intelligence, which only partially receives the full light of the sun, the universal intelligence, that is eternally present. As the moon alternates between proximity to and distance from the sun, so the lover alternates between the gain and the loss of the universal intelligence, even though it is, like the sun, eternally present. Here Bruno follows Ficino who compares the human soul to the moon's light and God to the light of the sun. God is the sun, which gives light to all things known by the human soul,

---

"Giordano Bruno's *De gli eroici furori* and the Emblematic Tradition", *Romanic Review*, LV: 1 (February, 1964), pp. 1-15.

[19] GIOVIO, pp. 20-21.

[20] ANDREA ALCIATI, *Emblematum flumen abundans*, ed. Henry Green (London, 1871), plate 53.

[21] THÉODORE DE BEZE, *Les Vrais Portraits des hommes et quarante-quatre emblèmes Chréstiennes* (n. p., 1581), emblème 6, p. 246.

even though the soul may ignore the origin of that light. [22] Bruno's emblem of a ship at sea is described as showing a young boy in a boat that is all but swallowed by an angry sea. In desperation the boy has abandoned the oars. The boy represents the lover who has momentarily lost faith in his ability to complete the voyage to the beloved object because of its overwhelming and incomprehensible excellence. [23] Bruno's emblem of a phoenix burning in the sun symbolizes the lover enraptured by the radiance of the divine ideal. As the phoenix sends up smoke which obscures the sun, so the lover only succeeds in obscuring the nature of the ideal by attempting to communicate it in verse. So incomprehensible is the nature of the divine being, that poets can never describe it adequately. Thus Bruno employs the emblem not only to symbolize the lover's difficulty in unraveling the enigma of his ideal, but to symbolize his characteristically negative theology as well. [24]

Bruno holds the divine being according to its own proper essence is persistently elusive, but its manifestation in nature need not be so. Accordingly, Bruno alludes to Diana, a symbol of the divine and infinite principle in nature in the final emblem of the series.

This emblem of a flaming yoke encircling a noose represents, as we are told in the commentary, the means by which Diana captures the lover, the felicitous bonds that have restricted the excesses of both the lover's ascetic and sensual desires, so that the purer elements of both might be set free. The emblem, therefore, suggests the lover in the ultimate phase of his progress, when Diana brings him the favor of the *summum bonum* on earth, the highest beatitude in this mortal state. [25]

---

[22] BRUNO, *Opere italiane*, 2 ed., 1, 5, 8, 408-411; my translation, pp. 159-161. See also MARSILIO FICINO, *Opera omnia* (Basel, 1561), 268-269, and AVERROES *De anima* (Venezia, 1550), iii, p. 165.

[23] BRUNO, *Opere*, 2, 1, 12, 459-462; my translation, pp. 213-215.

[24] *Ibid.*, 2, 1, 3, 436-440, my translation, pp. 187-191, 257 and note 23.

[25] *Ibid.*, 2, 2, 1, 463, my translation, pp. 216, 225-227, and *Opere*, 2, 5, 514-515, my translation, pp. 262-263. For Sir Philip Sidney's analogous allusion to the lover who aspires paradoxically to the "sweet yoke of lasting freedom", see son., "Leave me, O love, that reachest but to dust".

After Bruno, poets in the Renaissance continue to employ emblems in the total design of some of their poetic work. Francis Quarles in England is conspicuous among them. Indeed, before *De gli eroici furori*, Edmund Spenser had written *The Shepheardes Calender* in the emblematic tradition, a work which almost rivals that of Bruno in its extraordinary complexity. Even William Shakespeare and John Donne borrow often from the tradition those images which shape the underlying thought of some of their poetry.

After Bruno's *De gli eroici furori*, however, rarely, if at all, for reasons which perhaps suggest further inquiry, do we discover all three, emblems, sonnets and commentary combined so provocatively in a single work. More important, it is difficult, if not impossible, to find a work in which these are distinct, yet organically related parts of the author's total cosmological vision.

After Galileo's telescope confirmed the heliocentric theory, [26] which Bruno himself had helped to form, science and theology went their separate ways, and eventually became sharply antagonistic. The other disciplines followed suit, and, accordingly, philosophy divorced itself from poetry. Poetry, too, according to the new habit of mind, came to be increasingly isolated from the visual arts. *De gli eroici furori*, then, is one of the last works in the Renaissance in which all these serve to a remarkable degree as related elements of an organic whole.

---

[26] Galileo postulated a new physics of quantities that was based fundamentally, not on Aristotelian formal logic, but on mathematics. For further discussion of this matter see PAUL O. KRISTELLER, *Renaissance Thought* (Harper, 1961), pp. 44-47.

THE HEROIC FRENZIES

ARGUMENT OF THE NOLAN
UPON
THE HEROIC FRENZIES

Dedicated to the Most Illustrious Sir Philip Sidney

Most illustrious knight, it is indeed a base, ugly and contaminated wit that is constantly occupied and curiously obsessed with the beauty of a female body! What spectacle, oh good God, more vile and ignoble can be presented to a mind of clear sensibilities than a rational man afflicted, tormented, gloomy, melancholic, who becomes now hot, now cold and trembling, now pale, now flushed, now confused, or now resolute; one who spends most of his time and the choice fruits of his life letting fall drop by drop the elixir of his brain by putting into conceits and in writing, and sealing on public monuments those continual tortures, dire torments, those persuasive speeches, those laborious complaints and most bitter labours inevitable beneath the tyranny of an unworthy, witless, stupid and odoriferous foulness! [1]

What a tragicomedy! What act, I say, more worthy of pity and laughter can be presented to us upon this world's stage, in this scene of our consciousness, than of this host of individuals who became melancholy, meditative, unflinching, firm, faithful, lovers, devotees, admirers and slaves of a thing without trust-

---

[1] For sonnets written in parody of Petrarchist love poetry see FRANCESCO BERNI, *In lode della sua donna,* ed. Carli-Sarnati (Firenze, 1949), II, 165-166; William Shakespeare, Son. 130; see also Arturo Graf, "Petrarchismo ed anti-Petrarchismo", in *Attraverso il cinquecento* (Torino, 1888), pp. 1-86.

worthiness, a thing deprived of all constancy, destitute of any talent, vacant of any merit, without acknowledgment or any gratitude, as incapable of sensibility, intelligence or goodness, as a statue or image painted on a wall; a thing containing more haughtiness, arrogance, insolence, contumely, anger, scorn, hypocrisy, licentiousness, avarice, ingratitude and other ruinous vices, more poisons and instruments of death than could have issued from the box of Pandora? For such are the poisons which have only too commodious an abode in the brain of that monster! Here we have written down on paper, enclosed in books, placed before the eyes and sounded in the ear a noise, an uproar, a blast of symbols, of emblems, of mottoes, of epistles, of sonnets, of epigrams, of prolific notes, of excessive sweat, of life consumed, shrieks which deafen the stars, laments which reverberate in the caves of hell, tortures which affect living souls with stupor, sighs which make the gods swoon with compassion, and all this for those eyes, for those cheeks, for that breast, for that whiteness, for that vermilion, for that speech, for those teeth, for those lips, that hair, that dress, that robe, that glove, that slipper, that shoe, that reserve, that little smile, that wryness, that window-widow, that eclipsed sun, that scourge, that disgust, that stink, that tomb, that latrine, that menstruum, that carrion, that quartan ague, that excessive injury and distortion of nature, which with surface appearance, a shadow, a phantasm, a dream, a Circean enchantment put to the service of generation, deceives us as a species of beauty.

This is a beauty which comes and goes, is born and dies, blooms and decays; and is eternally beautiful for so very short a moment and within itself truly and lastingly contains a cargo, a store-house, an emporium, a market of all the filth, toxins and poisons which our step-mother nature is able to produce; who having collected that seed of which she makes use, often recompenses us by a stench, by repentance, by melancholy, by languor, by a pain in the head, by a sense of undoing, by many other calamities which are evident to everyone, so that one suffers bitterly, where formerly he suffered only a little.

But what am I doing? What am I thinking? Do I perhaps despise the sun? Do I regret perhaps my own and others having come into this world? Do I perhaps wish to restrict men from gathering the sweetest fruit which the garden of our earthly paradise can produce? Am I perhaps for impeding nature's holy institution? Must I attempt to withdraw myself or any other from the beloved sweet yoke which divine providence has placed about our necks? Have I perhaps to persuade myself and others that our predecessors were born for us, but that we were not born for our descendents? No, may God not desire that this thought should ever come into my head! In fact, I add, that for all the kingdoms and beatitudes which might ever be proposed or chosen for me, never was I so wise and good that there could come to me the desire to castrate myself or to become a eunuch. In fact I should be ashamed, whatever may be my appearance, if I should desire ever to be second to any one who worthily breaks bread in the service of nature and the blessed God. And that such participation can be of assistance to one's good intentions I leave for the consideration of him who can judge for himself. But I do not believe I am caught. [2] For I am certain that all the snares and nooses which those people devise and have devised who specialize in knotting snares and entanglements will never suffice for my enemies to ensnare and entangle me. They would avail themselves (if I dare say it) of death itself, in order to do me mischief. Nor do I believe myself to be frigid, for I do not think that the snows of Mt. Caucusus or Ripheus would suffice to cool my passion. See then if it is reason or some insufficiency which makes me speak.

What then do I mean? What conclusion do I wish to arrive at? What do I wish to decide? What I would conclude and say, oh illustrious knight, is that what belongs to Caesar be rendered unto Caesar and what belongs to God be rendered unto God. I mean that although there are cases when not even divine honors and adoration suffice for women, yet this does not mean that we *owe* them divine honors and worship. I desire that women should be honored and loved as women ought to be loved and honored. Loved and honored for such cause, I say, and for so much, and in the measure due for the little they are, at that time and occasion when they show the natural virtue peculiar to them.

---

[2] i. e., in a contradiction.

That natural virtue is the beauty, the splendor, and the humility without which one would esteem them to have been born in this world more vainly than a poisonous fungus occupying the earth to the detriment of better plants, more odious than any snake or viper which lifts its head from the dust. I mean that everything in the universe, in order that it have stability and constancy, has its own weight, number, order and measure, so that it may be ordered and governed with all justice and reason. Therefore Silenus, Bacchus, Pomona, Vertunnus, the god of Lampsacus and similar other gods of the drinking hall, gods of strong beer, and humble wine, do not sit in heaven to drink nectar and taste ambrosia at the banquet of Jove, Saturn, Pallas, Phoebus and similar gods; and their vestments, temples, sacrifices and rites must differ from those of the great gods.

Finally, I mean that these heroic frenzies have a heroic subject and object, and therefore can no more be esteemed as vulgar and physical loves than one can see dolphins in the trees of the forests or savage bears under the rocks of the sea.

However, to deliver all from such suspicion, I thought at first of giving to this book a title similar to the book of Solomon which under the guise of lovers and ordinary passions contains similarly divine and heroic frenzies, as the mystics and cabbalistic doctors interpret; I wished, in fact, to call it *Canticle*. But in the end I restrained myself for many reasons, of which I shall report but two. One for the fear which I conceived of the austere frown of certain pharisees, who would judge me profane for usurping sacred and supernatural titles in my natural and physical discourse, while they, consummate scoundrels, and ministers of every ribaldry, usurp more basely than one can say the names of holy ones, of saints, of divine preachers, of the sons of God, of priests, of kings. But then we await that divine judgment which will make manifest their malicious ignorance and doctrines; our simple liberty and their malicious rules, censures and institutions. The other for the great dissimilarity which is seen between the appearance of this work and that one, even though the same mystery and psychic substance is concealed under the shadow of the one and the other; for no one doubts that the first idea of the Sage [3] was to represent things divine rather than to present other things; with him the figure is openly and manifestly a figure, and the metaphorical sense is understood in such a way that it cannot be denied to be metaphorical, when you hear of those eyes of doves, that neck like a tower, that tongue of milk, that fragrance of incense, those teeth that seem a flock of sheep returning from the bath, those tresses that resemble goats descending the mountain of Galaad. But this poem does not show us a face which so keenly invites one to seek a latent and occult sense; so that through the ordinary mode of speech and by similitudes more adapted to the sentiments which gentle lovers usually employ, and experienced poets put in verse and rime, sentiments are expressed similar to those used by the poets who spoke of Cythereida, or Licoris, or Doris or Cynthia, Lesbia, Corynna, Laura and other such ladies. Thus anyone could be easily persuaded that my primary and fundamental intention may have been to express an ordinary love, which may have dictated certain conceits to me, and afterwards, because it had been rejected, may have borrowed wings for itself and become heroic; for it is possible to convert any fable, romance, dream and prophetic enigma, and to employ it by virtue of metaphor and allegorical disguise in such a way as to signify all that pleases him who is skillful at tugging at the sense, and is thus adept at making everything of everything, to follow the word of the profound Anaxagoras. [4] But think who will as it seems to him and pleases him, in the end, willy nilly, if one is to be just, each must understand and define it as I understand and define it, and not I as he would understand it and depict it; for just as the passions of that sage Hebrew have their own proper modes, succession and names, which no one has been able to understand and could never explain better than he, if he were present, so these canticles of mine have their own names, succession and modes which no one can explain better and understand than myself, since I am not absent.

---

[3] Solomon.
[4] SIMPLIKIOS *Physics* 156, 9; 164, 25; 157, 9; see also FELIX M. CLEVE, *The Philosophy of Anaxagoras* (New York, 1949), pp. 96-97.

Of one thing I wish the world to be assured: what I have essayed in this preliminary preface, wherein I address you in particular, excellent sir, and in the dialogues formed upon the subsequent articles, sonnets and stanzas, is to have everyone know that I should deem myself most shameful and bestial, if with much thought, study and labor I should have ever delighted or relished imitating (as they say) an Orpheus who adores a living woman,[5] and proposes after her death (if it be possible) to rescue her from hell; when in fact I would hardly esteem her (without blushing) to be worthy of being loved naturally even in that instant when her beauty is in flower and when she has the power of bringing offspring to nature and to God: so much the less would I desire to appear similar to certain poets and versifiers who glory in a perpetual perseverance in such love, as in such a pertinacious madness, which can certainly compete with all the other species of folly that can reside in a human brain. So much, I say, am I removed from that most vain, most vile and most infamous glory, that I cannot believe any man who possesses a grain of sense and spirit can expend any more love on such a thing than I have spent in the past and intend to spend in the present. And, by my faith, if I wish to employ myself in defending the nobility of that Tuscan poet, who showed himself so distraught on the banks of the Sorgue for a lady of Valclusa, and not say that he was a madman fit to be chained, I shall have to believe and force myself to persuade others, that for lack of genius apt for higher things he set himself the task of nourishing his melancholy, and belaboring his wit in confusion, by analyzing the effects of an obstinate vulgar love, animal and bestial,[6]

---

[5] Although Bruno tells us his sonnets are not intended for "a living woman", Vincenzo Spampanato suggests that the lady, Laodomia, of the family of Savolina in Nola, one of the two interlocutors of the closing dialogue, was for Bruno a Beatrice whose grace raised him to intellectual love. See below, p. 258 and note 1.

[6] This anti-Petrarchist tirade is not uncommon among those poets in the Renaissance who belong, in the final analysis, to the Petrarchan tradition which they satirize. Accordingly in this passage Bruno follows Boccaccio, Du Bellay, Wyatt and Sidney who distinguish sharply between the vulgar or bestial love of woman and the divine love of which she may also be the symbol. See GIOVANNI BOCCACCIO *Filostrato* 6, 34, and 8, 29-32; JOACHIM

as so many others have done who formerly have sung the praises of a fly, a beetle, an ass, of Silenus, of Priapus, of apes, and those who have in our time sung the praises of urinals, of the shepherd's pipe, of beans, of the bed, of lies, of dishonor, of the furnace, of the knife, of famine, and of the plague, things which perhaps give the appearance of being no less lofty and proud by reason of the celebrated voices of those who sing of them than these and other ladies I have mentioned are, perhaps by reason of the poets who have celebrated *them*. [7]

Yet (that there be no mistake) I do not wish that here should be taxed the dignity of those ladies who have been worthily praised and who are praiseworthy: and those, especially, who may and do reside in this British land, to whom we owe the love and fidelity of the guest; for even if one were to find fault with the whole world, one could not find fault with this nation, which in this respect is not the terrestrial world, nor a part of it, but is entirely separated from it, as you know: so that any discourse regarding the whole feminine sex could not and would not include any of your women, who must not be considered part of that sex; because they are not women, they are not ladies, but, in the guise of ladies, they are nymphs, goddesses and of celestial substance, among whom it is permitted to contemplate that unique Diana, whom I do not desire to name in the rank or category of women. [8] Let it be understood, then, that I mean only the ordinary genus. And I should unworthily and unjustly persecute any individual of this class: because to no particular person ought the weakness and condition of the sex be imputed, just as a defect or vice of constitution, assuming there is some fault or error there, must be attributed to the species or to nature, and not in particular to the individuals of the class. Truly, with respect to that sex, what I abominate is that zealous and disorder-

---

Du Bellay, son., "Contre les Petrarquistes", and *L'Olive*, 1; Sir Thomas Wyatt, son., "Farewell, love, and all thy laws forever"; Sir Philip Sidney, son., "Thou blind man's mark, thou fool's self-chosen snare", and "It is most true that eyes are formed to serve".

[7] i. e. by reason of the Latin poets alluded to on page 57, Petrarch and the Petrarchists, whom Bruno places among burlesques poets.

[8] Queen Elizabeth. See Giordano Bruno, *Opere italiane*, 2 ed. Gentile (Bari, 1925), II, p. 317 and note 2. See also above, introduction, pp. 44 and note 44.

ed venereal love which some are accustomed to expend for it, so
that they come to the point of making their wit the slave of
woman, and of degrading the noblest powers and actions of the
intellectual soul. If my intentions are understood, far from being
saddened and becoming vexed with me because of my natural
and truthful discourse, every honest and chaste woman will
rather agree with me and love me the more because of it; and
they will allow that the venereal love women have for men is a
dishonorable thing, as I actively reprove the venereal love men
have for women. Therefore, with a determined heart, mind,
opinion and purpose, I affirm that my first and principal, secondary and subordinate, final and ultimate design in this work to
which I have been called, was and is to signify divine contemplation and present the eye and ear with other frenzies, not
those caused by vulgar love, but those caused by heroic love.
These frenzies will be explained in two parts, each of which will
be divided into five dialogues.

THE ARGUMENT OF THE FIVE DIALOGUES OF THE FIRST PART

In the first dialogue of the first part there are five articles, [9]
whence, in order: in the first is shown the causes and principal
intrinsic motives under the names and figures of the mountain,
and the river, and of the muses which declare themselves present,
not because they have been summoned, invoked and searched
for, but rather as if they had often importunately offered themselves. By this is signified that the divine light is ever present,
that it forever offers itself, ever calls and knocks at the doors of
our senses and other powers of cognition and apprehension, as
it is indicated in the Song of Solomon where it is said, "En ipse
stat post parietem nostrum, respicinse per cancellos et prospiciens
per fenestras", [10] which light very often through various occasions

---

[9] i. e. the sonnets in each dialogue.

[10] Cant. 2:9 (Vulgate, Douay Version, 1609): "Behold He standeth
behind our wall, looking through the windows, looking through the lat-
tices..." Unless otherwise noted, all subsequent references to the Bible are
to the Douay Version.

and impediments remains excluded and withheld. In the second article is shown what are those subjects, objects, affections, instruments and effects by which this divine light enters, shows itself, and takes possession of the soul, in order to raise it and convert it unto God. In the third, the intention, definition and determination which the well-informed soul makes with regard to the one, perfect and ultimate end. In the fourth, the civil war which follows and breaks out against the spirit after such determination, whence the Canticle says, "Noli mirare, quia nigra sum: decoloravit enim me sol, quia fratres mei pugnaverunt contra me, quam posuerunt custodem in vineis". [11] In that place are represented as four standard bearers the affection, the fatal impulse, an appearance of the good, and the conscience, which are followed by the numberless cohorts of the many, contrary, varied and diverse powers, together with their ministers, intermediaries and organs which exist in this organization. In the fifth is described a natural contemplation through which it is shown that every contrary is reduced to friendship, whether through the victory of one of the contraries, or through harmony and conciliation, or by some vicissitude, every discord to concord, every diversity to unity; which doctrine has been developed by us in the discourses of other dialogues. [12]

In the second dialogue is more explicitly described the order and action of the conflict which is in the substance of this complex of the frenzied one, to wit: in the first article are shown three sorts of contraries. The first is the conflict of two opposed affections or acts, as for example where hopes are cold and desires hot. The second treats of the same desires and acts in themselves, not only at different times, but at the same time, when each one, for instance, dissatisfied with himself, looks to another, and at the same times loves and hates. The third is between the power that follows and aspires and the object which flees and eludes it. In the second article is described the opposition which results from

---

[11] Cant. 1:5: "Do not consider me that I am brown, for the sun has altered my color: for my brothers have fought against me, whom they have made the keeper in the vineyards..."

[12] GIORDANO BRUNO, *De la causa*, in *Opere italiane*, 2 ed. Gentile, I, 247-250.

two impulses which are opposed in general, to which are related
all the particular and subordinate contraries, for example, when
one climbs or descends toward two opposite places or goals at
the same time. Thus it happens to the complex being by reason
of the diversity of the inclinations which are in his several parts
and the variety of dispositions which result from these, that he
rises and falls at the same time, goes forward and backward,
withdraws himself from himself and also withdraws into himself.
The third article discusses the consequence of such oppositions.

In the third dialogue is disclosed how much power belongs
to the will in this combat, for to the will alone pertains the
organizing, the initiating, the execution and completion; for it is
the will the Canticle addresses when it says, *Surge, propera, co-
lumba mea, et veni: iam enim hiems transit, imber abiit, floras
apparuerunt in terra nostra; tempus putationis advenit.* [13] It is the
will that in many ways bestows power to the other potencies;
and bestows power especially to itself, when it reflects upon
itself and increases itself two-fold, when it wishes to desire, and
is pleased with what it desires; it withdraws itself, on the contrary, when it dislikes the object of its desire, and is displeased
to desire it. Thus everywhere and in everything it approves what
is good and what the justice of natural law prescribes for it, and
never approves at all what deviates from that law. And this is
how much the first and second article explain. In the third article
is seen the double fruit of a similar power. Accordingly, as the
result of the passion which draws and ravishes them, lofty things
become base, and base things become lofty. Thus it is customary
to say that by the force of vicissitude and vertiginous attraction,
the element of fire is condensed into air, vapor and water, while
water is refined into vapor, air and fire.

In the seven sections of the fourth dialogue are contemplated
the impetus and vigor of the intellect which carries the affection
away with it; the development of the thoughts into which the
frenzied lover is divided, and the sufferings of the soul under

---

[13] Cant. 2:10-12: "Arise, hasten, my dove, and come: for already
winter is passed, the rain is gone, the flowers have appeared in our land;
the time of pruning is come..."

the government of this so turbulent republic. There it becomes clear who the hunter is, the birdcatcher, the wild beast, the dogs, offspring, the cave, the noose, the rock, the prey, the issue of so many labors, peace, rest and the desired end of so laborious a conflict.

In the fifth dialogue is further described the state of the frenzied one and is shown the order, condition and reason for his labors and fortunes. In the first article is shown what pertains to the pursuit of the object which withdraws itself; [14] in the second is shown the continuous and relentless competition of the passions; in the third the lofty and cold, because vain purposes; in the fourth the voluntary desire; in the fifth the prompt rescue and powerful bulwark. In the following articles are shown in their variety, according to their reasons and appropriateness, the vicissitudes of his fortune, condition and labors, each article expressing them by antitheses, comparisons and similitudes.

ARGUMENT OF THE FIVE DIALOGUES OF THE SECOND PART

In the first dialogue of the second part is offered the origin of the modes and reasons for the state of the frenzied lover. In the first sonnet is described his state beneath the wheel of time; in the second is described the defense he offers for his esteem of ignoble occupations and for the unworthy squandering of time which is so brief and narrowly measured; in the third he confesses the impotence of his studies, which, although illumined within by the excellence of their object, begin to obscure and cloud that object when they come in contact with it; in the fourth he complains of the profitless strain of the faculties of the soul as his soul seeks to rise with powers unequal to the state it desires and venerates; in the fifth is recalled the contrariety and familiar conflict found in him, a conflict which may hinder him from applying himself entirely to his end or goal. In the sixth is expressed the aspiration of desire; in the seventh in considered the poor

---

[14] See below, p. 144 and note 3.

correspondence found between him who aspires, and that to which he aspires; in the eighth is seen the distraction the soul suffers because of the conflict between external and internal things, internal things among themselves, and a similar conflict of external things among themselves; in the ninth is explained the age and the time in the course of life most propitious for the act of lofty and profound contemplation, a time when the soul is not disturbed by the ebb and flow of its vegetative constitution, but finds itself in a state of immobility and in a sort of tranquillity; in the tenth is described the order and manner in which heroic love sometimes attacks, wounds and awakens us; in the eleventh is explained the multitude of species and particular ideas which show the excellence of the mark of their unique source and are the means by which the desire toward the heavenly is aroused; in the twelfth is expressed the state of every human effort toward the divine enterprises. Much is presumed before one engages himself in them, and much during the engagement itself. But, then, when one is engulfed and penetrates more and more into the depths, this fervent spirit becomes extinguished by presumption, the nerves begin to yield, the strength is slackened, thoughts discouraged, all intentions vanish, and the soul remains confused, vanquished and reduced to nothing. Pertinently, therefore, was it said by the Sage, *qui scrutator est maiestatis, opprimetur a gloria.* [15] In the last article is more clearly expressed what the twelfth demonstrated by similitude and figure.

In the second dialogue, in a sonnet and in the dialogue which is a commentary upon it, is made specific the first cause which subdued the strong one, softened the hard one, and reduced him to an amorous servitude under the command of Cupid, but in that way raised and disposed him to celebrate his zeal, ardor, election and purpose.

In the third dialogue in four questions and four answers of the heart to the eyes and the eyes to the heart is explained the being and mode of the appetitive and cognitive faculties. In this dialogue is shown how the will is reawakened from sleep, given

---

[15] Prov. 25:27: "...he that is a searcher of majesty shall be overwhelmed by glory..."

direction, urged and led by the cognition; and reciprocally how the cognition is aroused, formed and revived by the will, the one proceeding from the other, alternately. It is doubted if the intellect or the cognitive power in general, or even the act of cognition is greater than the will or appetitive power in general, or even greater than the affection. If one cannot love more than one can understand, and if everything which in a certain mode is desired, in a certain mode is also understood, and the reverse also be true; then it is fitting to call the appetite cognition. For we see that the doctrine of the Peripatetics, which has raised and nourished us from our youth, goes so far as to call the appetite in potency and natural act cognition, so that they distinguish all effects, means and ends, principles, causes and elements into those primarily, intermediately, and ultimately known according to nature, in which, they conclude, the appetite and the cognition concur. Thus is proposed the infinite potency of matter, and the assistance of the act thanks to which that potency is not in vain. [16] For just as the act of the will is infinite with respect to the good, so is the act of cognition infinite and endless with respect to the true: accordingly, being, truth and goodness take on the same significance when they are referred to in the same way, that is: as infinite goals.

In the fourth dialogue are represented and in some manner explained the nine reasons for the ineptitude, disproportion and deficiency of the human sight and apprehensive potency toward things divine. The first lover, who is blind from birth, is blind because of the nature which debases and humiliates him. The

---

[16] Although Bruno is conditioned strongly by Aristotelian terminology, he differs considerably from Aristotle's view of material potency. Aristotle conceives matter merely as abstractly potential. Bruno, however, follows Averroës and makes matter metaphysically subsistent and that which bears within itself the forms peculiar to it and brings them to realization in its movement. Thus he refers to "the infinite potency of matter" assisted by an equally infinite cognition", or, "act." See BRUNO, *De la causa*, in *Opere*, I, 238-239. For Aristotle's view that matter is the merely potential and has in itself alone no principle of motion or of generation see ARISTOTLE *Metaphysics* vii, 6, 1045b, 21. For Averroës' view of matter's metaphysical subsistence see MAURICE DE WULF, *History of Medieval Philosophy*, trans. P. Coffey (Longmans, Green, New York, 1909), p. 234.

second lover, blinded by the poison of jealousy, is blind because of the irascible and concupiscible which diverts and misleads him. The third, blinded by the sudden appearance of intense light, is blind because of the brilliance of the object which dazzles him. The fourth, raised and nourished for a long time in the light of the sun, is blind because of much lofty contemplation of the unity which removes him from the multitude. The fifth, whose eyes are forever filled with dense tears, is blind owing to the disproportion of means between the potency and the object which impedes him. The sixth, who through much weeping has extinguished the organic visual humour, is blind because of a lack of the true intellectual nourishment, a lack which weakens him. The seventh, whose eyes are reduced to ashes by the ardor of his heart, symbolizes the burning passion which disperses, weakens and sometimes devours the power of discernment. The eighth, blinded by the wound of an arrow's point, is blind through the very act of union with the form of the object that conquers, alters and seduces the apprehensive potency, which is oppressed by the weight of the form and falls under the impetus of its presence; therefore, not without reason is the appearance of this object sometimes represented in the form of a penetrating thunderbolt. [17] The ninth, because he is mute and is unable to explain the cause of his blindness, is blind for the highest reason, the secret design of God, who has given man this zeal and solicitude to search, so that he may never be able to reach higher than to the knowledge of his own blindness and ignorance, and no higher than to deem silence more worthy than speech. But this does not suggest that common ignorance is to be excused or favored, for he is doubly blind who does not see his own blindness. And there is a difference between the profitably zealous and the stupidly idle. The stupidly idle are buried in the lethargy of the incapability of

---

[17] PETRARCH *Rime* 133:

> ...L'Amor m'a posto come segno a strale...
> I penser son saette, e 'l viso un sole,
> e 'l desir foco; e 'nsieme con quest' arme
> Mi punge Amor, m'abbaglia, e mi distrugge...

See below, p. 87 an note 10, and p. 88.

judging their own blindness, and the profitably zealous are aware, awakened and prudent judges or their own blindness, and for that reason are in quest and on the threshold of the attainment of the light from which the others are banished for a long time.

ARGUMENT AND ALLEGORY OF THE FIFTH DIALOGUE

In the fifth dialogue two women are introduced, for whom (according to my country's custom) it is unbecoming to comment, expound, decipher, or to be so wise and learned as to usurp the office of teaching and giving men institutions, rules and doctrines, but for whom it is fitting, when their bodies are found to have a soul, to divine well and to prophecy. Therefore the author has been content to make them merely recite the allegory, leaving to some male intelligence the care and labor of interpreting it. And even to him (in order to lighten his task, or I should say, discharge him of it), I shall explain how these nine blind men, by reason of their role, of the external causes of their blindness and of many other subjective differences, take on significance other than the nine of the preceding dialogue. According to the common imagination of the nine celestial spheres these blind men symbolize the number, order and diversity of all things which are subsistent within an absolute unity, and in and over all of them are ordered those intelligences which, by a certain analogy, depend upon the first and the unique intelligence. [18] The Cabalists, Chaldeans, Magi, the Platonists and Christian theologians hold that these intelligences are distinct in nine orders through the perfection of the number which governs the universality of things and in a certain way informs everything. They also hold that it is by a simple number that the divinity is symbolized, whose extension and square represents the number and substance of all things which depend upon it. All the more illustrious thinkers, whether

---

[18] Bruno follows Dante in the *Vita nuova* and employs mystic numerology for the design of the closing dialogues of *De gli eroici furori*. Accordingly the nine lovers, first blind, are illuminated by the divine power of the "simple number", the root of nine, three, a trinity. See above, introduction, p. 26 and note 15.

philosophers or theologians, who speak either by reason and
their own light, or by faith and a superior light, recognized in
these intelligences the cycles of ascent and descent. Thus the
Platonists say that by a certain revolution it happens that those
who are above the fatality of time and change submit themselves
once again to this fatality, while others rise and take their place.
A similar revolution is alluded to by the Pythagorean poet, when
he says:

> Has omnes, ubi mille rotam volvere per annos
> Lethaeum ad fluvium deus evocat agmine magno,
> Rursus ut incipiant in corpora velle reverti. [19]

Some say that thus are to be understood the words of Revelation in which it is said that the dragon shall be conquered by
chains for a thousand years, and after that period released. [20] To
this interpretation adhere those who speculate upon the many
passages of Revelation which express the millenium literally, represent it by a year, by a season, by one night, or by one span
time or another. Beyond a doubt the millenium itself is not to
be taken according to the revolutions called solar years, but
according to more than one method of calculating the order and
measure upon which the fate of things depends. For the years of
the stars are as different as are their particular species. As for the
fact of revolution, it is given out among the Christian theologians
that from each of the nine orders of spirits, a multitude of legions
were cast down to low and obscure regions; and so that those
seats do not remain vacant, divine Providence wishes the spirits
who now live in human bodies to be drawn up to that eminence. [21]
But among the philosophers Plotinus alone, to my knowledge,
has seen fit to agree with all the great theologians that such

---

[19] VIRGIL Aeneid vi. 748-751:

> All these, where the wheel of a thousand years comes round, a
> god summons to the river Lethe in vast train, so that they may
> begin again to desire the return to the body...

[20] Apoc. 20:2, 7.
[21] ST. AUGUSTINE, De civitate, 22, 1, P. L., t. 51, col. 751.

a revolution does not concern all beings, nor take place at all times, but takes place only once. And among the theologians only Origen, [22] following all the great philosophers, has dared to say, after the Saducees and other reproved sects, that the revolution is vicissitudinal and yet eternal, and that all those who ascend must descend to the bottom; as one can see in all the elements, and in all the things which exist on the surface, in the bosom and womb of nature. For my part, I confess and confirm as very appropriate the opinion of the theologians and those whose task it is to give laws and institutions to the people; just as I do not fail to affirm and accept the opinion of those who, speaking according to natural reason, address themselves to the small number of the good and wise. The latter opinion has been justifiably reproved for having been exposed to the eyes of the multitude, for since it is only with great difficulty that they can be restrained from vices and spurred to virtuous action by belief in eternal punishment, what would happen were they persuaded of some lighter condition for the reward of heroic and human deeds, and the punishment of crimes and villainies? But to conclude this progression of mine, I say that now begins an explanation and discourse upon the blindness and the light of these nine men, first clairvoyant, then blind, and finally illumined. At first they are rivals in the shadows and vestiges of the divine beauty; then they are completely blind, and finally they enjoy themselves peacefully in the more open light. While they are in the first condition, they are led to the dwelling of Circe, who represents the generative matter of all things. She is called the daughter of the sun, because from the father of forms she has inherited the possession of all those forms which, by a sprinkling of the waters—that is to say by the act of generation and by the power of enchantment—that is by reason of a secret harmony—she transforms all beings, making those who see become blind. For generation and corruption are causes of oblivion and of blindness, as the ancients explain by the figure of souls who bathe and inebriate themselves in the waters of Lethe. [23]

---

[22] Origen, however, in *De princ.* III, 1, 21, P. G., t. 11, col. 17 believes these vicissitudes are not eternal, but end at the consummation of the world.
[23] Below, p. 261 and note 6.

Then by that which the blind men lament, when they say, *Daughter and mother of darkness and horror,* is signified the dismay and sadness of the soul which has lost its wings, but will be relieved when it regains hope of recovering them. By Circe's words, *Take another one of my fatal vases,* is signified that men carry with themselves the decree and destiny of a new metamorphoses, which is, however, said to be offered to them by Circe herself; for although one contrary has its origin in the other, it may not be efficaciously uncovered by them. For that reason she said that although her own hand was unable to open it, it could entrust the vase to them. The other meaning is that there are two kinds of water. There are the inferior waters under the firmament which enlighten. [24] These are the waters which the Pythagoreans and the Platonists symbolized by the descent from one tropic and the ascent to another. Then by her words, *Traverse the width and depth of the world, seek out all the many kingdoms,* is signified that there is no immediate progress from one contrary form to another, nor immediate regression to the first form, but that it is necessary to traverse, of not all, at least a very great number of the forms contained in the wheel of natural species. Then will they be enlightened by the sight of the object in which concur the three perfections, beauty, wisdom and truth, revealed through the sprinkling of the waters, called in the sacred books the waters of wisdom and the rivers of eternal life. These waters are not found on the mainland of the globe, but *penitus toto divisim ab orbe,* [25] in the bosom of the Ocean, of the Amphitrite, of the divinity, where that river rises which takes its source from the divine throne, whose flow is not at all like the ordinary flow of natural rivers. In that river are the nymphs, who are the blessed and divine intelligences which assist and administer to the first intelligence, similar to Diana among the nymphs of the wilderness. She alone among all the others has by her triple virtue the power to open every seal, untie every knot, uncover every

---

[24] Gen. 1:6, 7. See also ORIGEN, *Homilies upon Genesis,* I, 2 (Christian sources), p. 66, and ST. JEROME, *Epist.* 51, 5, who opposes the allegorical interpretation of the verses in Genesis.

[25] "...separated entirely from the earth..."

secret and bring to light whatever is hidden. By her unique presence, by her double splendor of goodness and truth, benevolence
and beauty, she pleases all wills and intellects, sprinkling them
with the salutary waters of purgation. Then there follows a long
chant and song by the nine intelligences, the nine muses, whose
chorus is ordered according to the number of the nine spheres,
so that the harmony of each one is continued by the harmony of
the following one. And that there may be no vacuum interposed
among them, the end of one song coincides with the beginning of
the other, and the end of the last song concurs with the beginning
of the first, as the circle is closed. [26] For the most brilliant and
the most obscure, the beginning and the end, the greatest light
and the most profound darkness, infinite potency and infinite act
coincide, as our method of argument has explained elsewhere. [27]

Finally one observes the harmony and concert of all the
spheres, intelligences and muses in a consort of instruments, so
that the heaven, the movement of worlds, the works of nature,
the discourse of intellects, the contemplation of the mind, the
decree of divine Providence celebrate in complete accord that
lofty and magnificent vicissitude which raises the inferior to the
superior waters, changes night into day, and day into night, so
that the divinity may be in all, according to the mode in which
the infinite goodness is infinitely communicated according to the
entire capacity of each thing.

These are the discourses, then, which it seems to me cannot
be conveniently addressed and recommended to anyone than to
you, excellent Sir. For I would not risk doing again what I think
at times I have done inadvertently, and what many others ordinarily do who present a lyre to a deaf man and a mirror to a blind
one. To you then these discourses are presented without fear,
because here the Italian reasons with one who understands him.
My verses are submitted to the censure and the protection of a
poet. My philosophy stands naked before so pure an intellect
as yours. [28] Heroic things are addressed to the heroic and gener-

---

[26] See below, pp. 263-267.
[27] BRUNO, De la causa, in Opere, I, 263-264.
[28] Bruno in the De immenso celebrates similarly his "philosophia nuda",
whose body is the source of light: "Nudaque illa est... nudaque de toto

ous spirit with which you are endowed. My services are offered to one who knows how to accept them graciously, and my homage to a gentleman who has ever shown himself worthy of such. And in that which particularly concerns me, I know that through your good services you have guided me with a magnanimity far greater than any recognition you may have given to others who may have since come to you. Farewell.

iaculatur corpore lucem". See *De immenso*, viii, 1, 86-87, and GENTILE, *Giordano Bruno e il pensiero del Rinascimento* (1925), 191-192.

The Apology of the Nolan

To the most glorious and virtuous ladies

Oh glorious and enchanting nymphs of England,
my spirit neither shuns nor disdains you, nor dishonors
you when it deprives you of the traditional name of
women,

by neither counting you among them nor excluding you.
I am sure the name of goddesses are more meet for you,
because you are endowed with more than common life,
and are upon the earth what the stars are in heaven. [1]

Oh, ladies mine, your sovereign beauty my sincerity
can never harm, nor does it wish to do so, because it
cannot reach your superhuman kind,

but by bitter torment, it aspires to that place
where Diana is queen above all, who is among you
what the sun is amidst the stars. [2]

Labor and art humbly offer you my invention, my
words and the strokes of my pen such as they may be.

---

[1] See above, "Argument of the Nolan", p. 65.

[2] PETRARCH *Rime* 9:

> così costei, ch'è tra le donne un Sole,
> In me, movendo de' begli occhi i rai...

> thus she, who is a sun among women,
> moving, within me, the rays of her beautiful eyes.

PETRARCH *Trionfi* ii. 44:

> E veramente è fra le stelle un sole...

> And truly is she a sun amidst the stars...

FIRST PART OF THE HEROIC FRENZIES

FIRST DIALOGUE

INTERLOCUTORS

TANSILLO   CICADA [1]

TANSILLO   The frenzies, then, most worthy of being placed in
the first rank and considered first are those I present to
you in the order that has seemed to me most convenient.

CICADA   Begin to read them then.

TANSILLO   Muses, whom I have so often rejected, importunate
cohorts of my suffering, alone consoling me in my
woes by such verses, rimes and frenzies

the like of which you never showed to others who
boast of the myrtle and the laurel; now let the wind,
anchor and port keep me close to you, if I am
forbidden to cruise elsewhere.

---

[1] Luigi Tansillo, Neapolitan poet (b. 1510, d. 1568), was of a noble
but poor family and fought for Spain against the Turks. Under the patronage
of the Spanish Viceroy, Pietro de Toledo, he composed his *Poesie liriche* to
a lady believed to have been the wife of the Marchesa del Vasto, Marie of
Aragon, who scorned his love. Four sonnets of the *Poesie liriche* appear in
De gli eroici furori.

Odoardo Cicada was, in all probability, a friend of Giovanni Bruno, father
of Giordano Bruno, and served as a soldier under Philip II until 1598. See
VINCENZO SPAMPANATO, *Vita di Giordano Bruno* (Messina, 1921), p. 65 and
note 2; see also BRUNO, in *Opere italiane*, II, 331 and note 2.

Oh mountains, oh goddesses, oh streams, where I
live, converse and nourish myself; where I learn in
quiet and find beauty;

through whom I rise, reawaken, adorn my heart,
[spirit
and brow; may you transform death, cypresses and
infernos into life, into laurels, into eternal stars.

One may infer that he rejected the muses often
and for many reasons, among which perhaps are these.
First, because he was not able to be idle, as the priest
of the muses must be; for one cannot be idle who must
defend himself against the ministers and servants of
envy, ignorance and malice. Second, because he had
received no assistance from worthy protectors and
defenders, who might have given him security. As it
is said by the poet:

Oh Flaccus, there will be no want for Maros,
if there is no lack of Maecenae. [2]

Another reason was that he regarded himself obligated to devote himself to the contemplation and
philosophical studies, which if not more advanced in
maturity, ought none the less, as mothers to the Muses,
to come before them. Moreover, because the tragic
Melpomene drew him on the one hand with more
matter than talent, and the comic Thalia drew him on
the other hand with more talent than matter, it happened that as one took from the other, he stood between the two weak and idle, rather than doubly active.
Besides, he had become a victim of the authority of
the censors, who, turning him from the more worthy
and noble things to which he was naturally inclined,
shackled his intellect, in order to enslave him beneath
the rule of a most vile and senseless hypocrisy, from
the freedom he had under the rule of virtue. But
finally, because of the great heat of annoyance into
which he fell, it happened that having nothing else

---

[2] MARTIAL *Epigram* 8:

Sint Maecenates, non deerunt,
Flacce, Marones.

from which to draw consolation, he accepted the call of those who are said to have inspired him with certain frenzies, verses and rimes, the like of which they never shared with anyone else. It is for that reason that this work sparkles with originality more than with imitation.

CIC.   Tell me, what is meant by those who praise themselves by means of the myrtle and the laurel?

TANS.   Those who can and do win praise for themselves by the myrtle are those who sing of love. If these bear themselves nobly, they win the crown of that plant consecrated to Venus who inspires them with her frenzy. Those who can praise themselves by the laurel are those who sing worthily of heroic things, who instruct heroic souls through speculative and moral philosophy, or who celebrate those heroic souls and present them as exemplary mirrors of political and civil action.

CIC.   Are there still other species, then, of poets and awards?

TANS.   There are not only as many as there are Muses, but a great many more besides. For, although one can distinguish certain sorts of poets and awards, one would not know how to define certain modes and species of human genius.

CIC.   I know certain makers of poetic rules who accept with difficulty Homer as a poet, and who reject Virgil, Ovid, Martial, Hesiod, Lucretius and many other versifiers, after having examined them according to the rules of Aristotle's *Poetics*.

TANS.   You can be sure, my friend, that these are veritable blockheads, for they do not consider that those rules serve chiefly to make clear the nature of the poetry of Homer, or the nature of some other particular poet. They do not consider that those rules are there only to show us the kind of epic poet Homer was, and not to serve as modes of instruction to other poets who could in other veins, skills, and frenzies be in their several kinds equal, similar or even greater than Homer.

CIC.   If I understand you correctly, then, Homer in his genre was not a poet who depended upon rules, but he is the cause of the rules which serve others who are more adept at imitating than inventing. And these rules were drawn up by an author who was not a poet of any sort, but who knew how to assemble rules of that particular kind (that is, rules of Homeric poetry) for the benefit of one who would wish to be not another poet with a muse of his own, but an imitator of Homer and the ape of Homer's muse.

TANS. You conclude well that poetry is not born of the rules, except by the merest chance, but that the rules derive from the poetry. For that reason there are as many genres and species of true rules as there are of true poets.

CIC. How will the true poets, then, be recognized?

TANS. By our singing their verses, and by this, that when they are sung, either they will be delightful, or they will be useful, or they will be useful and delightful at the same time. [3]

CIC. Whom then do the rules of Aristotle serve?

TANS. Those who cannot, as Homer, Hesiod, Orpheus and others could, be a poet without the aid of Aristotle. And they serve him who, not having a muse of his own, prefers to court the muse of Homer.

CIC. Then certain dismal pedants of our own day are wrong, who exclude some from the rank of poets because they do not conform their speech and metaphors or the introductions of their books and songs to those of Homer or Virgil, or because they do not observe the traditional use of the invocation, or because they entwine one story with another, or end their songs with summaries of what has been said already, and with announcements of what is to come; and because of other reasons drawn from a thousand methods of examination, of censures and rules in virtue of that text. [4] Therefore it appears that they themselves would be the true poets (should they so decide), and would easily attain the end toward which the others tend only with effort. But, if the truth were known, these pedants are nothing but worms, who do not know how to do anything well, and are born only to gnaw, soil and

---

[3] HORACE *Epistolarum* ii. 3, 333-334.

[4] Italian literary theorists in the sixteenth century considered Aristotle's *Poetics* an authoritative text from which it was not allowed to depart. Bernardo Segni, Florence, 1549, de Castelvetro, Vienna, 1570, d'Alessandro Piccolomini, Siena, 1571, and de Riccoboni, Venice, 1584, translated the *Poetics* in the vulgar tongue. During the same period Francesco Robortello, Florence, 1548, Maggi, Venice, 1550 and Pier Vettori, Florence, 1560 wrote Latin commentaries upon the same text.

hurl their dung upon the studies and labors of others; and being incapable of becoming illustrious through their own virtue and talent, they seek to advance themselves through the vices and errors of others.

TANS.   Now to return to the point from which passion has led us to digress to some extent, I say that there are and can be so many kinds of sentiment and human creations, which one can adorn with garlands not only of all sorts and species of plants, but also of all types and species of material. As a result, crowns for poets are made not only of myrtle and laurel, but also of the vine branch for scurrilous verses, of ivy for Bacchic verses, of olive for sacrifices and laws, of the poplar, elm and corn for agriculture, of cypress for funerals, and other garlands without number for as many other occasions; and, if you will permit, even of that material which a gallant gentleman designated, when he said:

> Oh Brother Porro, poet of flukes,
> at Milan you girdle yourself with a garland
> of pudding, tripe and sausage. [5]

CIC.   Therefore, through various talents which he displays in various meanings and purposes, this poet certainly will be able to adorn himself with branches of various plants, and be able to speak worthily with the muses, because near them he finds the air which comforts him, the anchor which sustains him and the port that welcomes him in time of fatigue, turmoil and tempest. Thus he says, Oh *mount* Parnassus *where I live, Muses* with whom *I converse,* stream of Helicon (or some other) *where I nourish myself,* mount which gives me tranquil abode, Muses who inspire me with profound doctrine, font which refreshes me and cleanses me of every stain, mount where I lift up my heart as I ascend, Muses conversing with whom I revive my spirit, font reposing under whose shadows I adorn my brow — change my death into life, my cypresses into laurels, and my infernos into heaven. That is to say, destine me to immortality, make me a poet, render me illustrious, the while I sing of death, cypresses and infernos.

TANS.   Good. Because for those who are favored by heaven, the greatest evils are converted into even greater good;

---

[5] See PIETRO ARETINO, *Capitoli, All' Albicante,* 16-18.

for necessity nourishes labors and studies, and these as a rule nourish the glory of immortal splendor. And so the death of one century brings life to all the others. [6]

CIC.   Continue.

TANS.   Next he says:

My heart is in the place and form of
Parnassus, which I must ascend for my safety; my
muses are the thoughts which at every hour reveal
to me their glorious tale;

my fount of Helicon is there, where my eyes
often pour forth profuse tears. Through such
mountains, through such nymphs and waters, as it
pleased heaven, I was born a poet.

Now let no king or favorable hand of
any emperor, or highest priest and sovereign
shepherd

give me such favors, honors and privileges.
My heart, my thoughts and my tears themselves cause
[the
laurel to bear leaves for my adornment.

Here first he declares what his mount is, speaking of it as the lofty passion of his heart; secondly, what his muses are, speaking of them as the beauties and prerogatives of his object; third, what his founts are, and these he speaks of as his tears. Upon that mount his passion is enkindled, out of beauties proceeds his frenzy, and by these tears is made manifest his passion.

In this way he deems himself no less able to be crowned illustriously through his own heart, thoughts and tears, than others who are crowned by the hands of kings, emperors and popes.   .

CIC.   Make clear to me what he means when he speaks of the heart in the form of Parnassus.

---

[6] For an analogous allusion to the cyclic theory of history, see ANGELO LIPARI, *Dolce Stil Nuovo according to Lorenzo de Medici* (Yale, 1936), pp. 90-91, 120-121. See also MACHIAVELLI's theory of political cycles in *I discorsi*, I, 2.

TANS.   By these words he means that the human heart
contains two summits, which rise progressively from
one root; and in the spiritual sense, from a single passion of the heart proceed the two contraries of hate
and love. [7] For Mount Parnassus has two summits
rising from the one foundation. [8]

CIC.   Continue.

TANS.   He says:

The Captain summons all his warriors
beneath a banner by the sound of the trumpet;
where, if it happens that for some of them it
sounds in vain, and they come not promptly,

those who are traitors he kills, the madmen
he banishes from his camp or he scorns them: so
the soul with those of its intentions which
come not to assemble under one standard, either it
wishes them dead or removed. [9]

I regard one object, which absorbs my mind, and
it is a single visage. I remain fixed upon
one beauty,

---

[7]   See BRUNO, *De la causa*, in *Opere*, I, 263 for the *coincidentia oppositorum* in psychological phenomena.

[8]   LUCAN *Pharsalia* v. 72-73:

*Parnassus gemino petit aethera colle
Mons Phoebo Bromioque sacer...*

[9]   DANTE *Vita nuova* 14:

... ...
*Ch'Amor, quando sì presso a voi mi trova,
Prende baldanza a tante sicurtate,*

*Che fiere tra' miei spirti paurosi
E quale ancide, e qual caccia di fuora,
Sicch' ei solo rimane a veder vui...*

... ...
*For love, when he finds me so near unto you,
Exultant grows and takes such assurance,*

*that he smites among my afflicted senses and
some he slays, and others he chases forth so
that he alone remains to gaze upon you...*

which has so pierced my heart, and is a single
dart; by one flame only I burn, and know but
a single paradise. [10]

The captain is the human will which sits at the stern
of the soul and with the little rudder of reason governs
the affections of the inferior potencies against the surge
of their natural violence. With the sound of the trumpet, that is to say, by determined election, he summons
all his warriors; that is, he calls forth all the potencies
of the soul (warriors we call them because they are in
continuous conflict and opposition), or the effects of
those potencies, which are the conflicting thoughts,
some of which incline toward one, and others toward
the other contrary; and he seeks to assemble them
beneath a single banner for a determined end. If it
happens that some of these thoughts which are required
to present themselves promptly and obediently are
called in vain, (especially those which proceed from the
natural powers that either do not obey the reason at
all or obey it very little), the captain is forced at least
to prevent those thoughts from taking action, and if
this cannot be accomplished, he condemns them; it is
thus that he is shown as one who would put some of
them to death and banish the others, proceeding
against the former with the sword of anger, and against
the latter with the whip of disdain.

Here *he regards one object* to which he is turned
by his intention. A *single visage pleases* him and
absorbs his mind. In a *single beauty* he is delighted
and pleased, and is said *to remain fixed upon it*, because the work of the intelligence is not an operation
of motion, but one of rest. And from that beauty only
does he conceive the *dart* which kills him; that is,
which summons him to the ultimate end of perfection.
He burns by one flame only, that is, he is sweetly
consumed by a single love.

CIC.   Why is love symbolized by fire?

---

[10] PETRARCH *Rime* 133:

Amor m' a posto come segno a strale
... ...
Da gli occhi vostri uscìo 'l colpo mortale

TANS.   Putting aside many other reasons for the moment,
let this suffice for you now. Love converts the thing
loved into the lover, as the fire, among all the most
active elements, is able to convert all the other simple
and complex elements into itself.

CIC.   Now continue.

TANS.   *He knows a paradise,* that is, a principal end; because paradise commonly means the end; and here
one must distinguish between the end which is absolute
in truth and essence, and that end which is so by similitude, shadow and participation. [11] According to the
first mode, there cannot be more than one end, just
as there is only one ultimate and prime good; according to the second mode, there are an infinite number.

Love, fate, the object and jealousy [12] are
for me pleasure, torment, content and distress.

... ...
Da voi sola procede, e parvi un gioco,
Il sole, e 'l foco, e 'l vento, ond' io son tale.

I pensier son saette, e 'l viso un sole,
E 'l desir foco; e 'nseme con quest' arme
Mi punge Amor, m'abbaglia, e mi distrugge;

E 'l 'angelico canto, e le parole,
Col dolce spirto, ond' io non posso aitarme,
Son l'aura inanzi a cui mia vita fugge.

... ...
From you alone proceeds the sun, and the fire,
and the wind, and seems game to you, wherefore
such am I.

Your thoughts are darts, and your face a sun,
and desire a fire; and with such armor love at once
strikes me, dazzles me, and destroys me;

And the angelic song, and words, with the sweet
spirit from which I cannot help myself, fills
the air in whose presence my life flees.

[11] PLATO *Phaedrus* 250: "There is no light of justice or temperance or
any of the higher ideas which are precious to souls in the earthly copies of
them: they are seen through a glass dimly; and there are few who going to
the images behold in them the realities, and of these only with difficulty..."
[12] Jealousy: Whereas the fate of the frenzied one is the uncertain distance
between his mind and the object he desires, jealousy is conceived here as the

The senseless boy, the blind and guilty one,
the supreme beauty and my one sole death

shows me paradise, and snatches it away,
presents me with every good, and withdraws it
from me; so much so that the heart, mind, spirit,
and soul have joy, have discomfort, have refresh-
ment and a heavy burden.

Who will rescue me from the conflict?
Who will make me enjoy the fruit of my good
in peace?

Who will put that which wearies me far
from that which delights me, so as to cause my
ardors and my tears to become happy ones?

In this verse he shows the cause and the origin
whence his frenzy is conceived and his enthusiasm is
born — by ploughing the field of the Muses, by scatter-
ing the seeds of his thoughts there, by aspiring to love's
harvest, and discovering the fervor of the sun in the
heat of his own passions and the humour of the rain
in his own tears. He places four things first: *love, his
fate, the object and jealousy.* Here love is not a base,
ignoble and unworthy mover, but a heroic lord and
his guide. Fate is nothing else than the fatal disposition
and order of mishaps to which he is subjected by his
destiny. The object is the lovable thing and the correl-
ative of the lover, and it is clear that jealousy is the
zeal of the lover concerning the thing loved; it is not
necessary to explain this to him who has tasted love,
and in vain shall we strain ourselves to explain it to
others. [13] *Love pleases* because to him who loves it is
pleasant to love; and he who truly loves would not
wish not to love. Wherefore I do not wish to omit
referring to that which I have shown in this sonnet
of mine: [14]

---

lover's pain in deprivation of the object. See above, p. 67; also, ANDREAS
CAPELLANUS, *De amore*, trans. John J. Parry (New York, 1941), pp. 45, 46,
102, 158, 185, and LORENZO DE MEDICI, *Commento sopra alcuni dei suoi
sonetti*, in *Opere* (Bari, 1913), II, p. 93.

[13] DANTE *Vita nuova* 14 (the *ragione*).
[14] LUIGI TANSILLO, *Poesia liriche*, ed. Fiorentino (Napoli, 1882), p. 15.

Dear, gentle and revered wound of that sweet
dart, which love ever chooses; lofty, gracious and
precious ardor, which makes the soul toss in ever
burning delight,

what virtue of herb, or force of magic art,
will ever release you from the center of my heart,
since the fresh onslaught which strikes there at
every hour, delights me the more it torments me?

My sweet pain, new in the world and rare,
when shall I ever escape from your burden, since
the remedy is weariness to me, and the pain delight?

Eyes, flames and bow of my lord, two-fold
fire in the soul, and arrows in the heart, because
the languishing is sweet to me, and the fire is dear.

His *fate torments* because of the unhappy and un-
wished for events, or because it causes the subject to
be esteemed less worthy of enjoying its object, and less
proportioned to its dignity; or because it does not permit reciprocal relation between the lover and his object; or for other reasons and obstacles which confront
him. The object makes the subject *content*, who does
not nourish himself with anything else, who seeks
nothing else, occupies himself with nothing else and
because of that object banishes every other thought.
Jealousy *distresses* inasmuch as it is the daughter of
that love from which it derives, the inseparable companion and sign of that love, —and where love manifests itself jealousy is understood as a necessary consequence, a counter-proof of which one can find among
generations which, from the frigidity of the climate
and backwardness of spirit, comprehend less, love little
and thus know nothing of jealousy— inasmuch, I say,
as it is the daughter of love, its companion and its sign,
it never ceases to disturb and poison everything found
beautiful and good in love. Therefore as I have said in
another one of my sonnets: [15]

Oh daughter so guilty of love and envy,
that you turn the joys of your father into pain, the

---

[15] *Ibid.*, p. 17.

adroit Argus [16] to disaster, and the blind idiot to
well being, minister of torment, Jealousy,

infernal Tisiphone, [17] fetid harpy, who seizes
and poisons the sweets of others; cruel Auster, [18]
through whom the loveliest flower of my hope must
languish;

wild beast odious to yourself, bird foreboding
of nothing but mourning, pain which enters the heart
through a thousand gates,

if one could deny you entrance, the kingdom of
love would be as sweet as a world without hate
and without death.

Add to what has been said that Jealousy is not only
sometimes the death and ruin of the lover, but on many
occasions kills love itself, especially when it nurtures
contempt; for then jealousy becomes so dominated by
its offspring that it extinguishes love and puts the
object to scorn; in fact, makes it no longer the object.

CIC.   Now explain the other particulars which follow;
that is, the reason why love is called *the senseless boy*.

TANS.   I shall explain everything. Love is called *the senseless boy*, not because it is foolish of itself, but because
it makes most lovers foolish and in such lovers is a
foolish thing. But in those who are the more intellectual
and speculative, love raises the mind the more and
purifies the intellect the more, awakening it, filling it
with zeal and prudence, developing a heroic ardor of
the soul, and an emulation of virtue and magnanimity
in the desire to please and become worthy of the thing
loved. By the majority love is understood as crazy and
stupid, for love makes most men pour forth their
peculiar sentiments and urges them on in exaggera-
tions, because it finds their spirit, soul and body badly
constituted and incapable of considering and distinguishing what is fitting for them from what renders

16   Ovid *Metamorphoses* i, 621-721.
17   *Ibid.*, iv, 480-530.
18   *Ibid.*, i, 262-274.

them more deformed, and thus makes them subjects of scorn, laughter and vituperation. [19]

CIC.

They say commonly and proverbially that love makes old men mad, and young men sages.

TANS.

The former unseemliness does not fall to all old men, nor does the latter advantage fall to all young men; but it is true of the latter who are well constituted, and of the former who are badly constituted. And therefore it is certain that whoever is accustomed in youth to love with discernment, in old age will love without going astray. But derision and laughter belong to those who at a mature age would, as it were, begin to learn their alphabet.

CIC.

Now tell me, why is his destiny or fate called *blind* and guilty?

TANS.

Fate is called *blind* and even *guilty* not of itself, for it is the very number and measured order of the universe; but with respect to its subjects it is called blind and is blind because it renders them blind to its view by being itself most uncertain. And similarly fate is called guilty because there is no mortal whose lamentations and complaints do not accuse it in some way. Thus the Apulian poet said:

> How is it Maecenas, that no one in the world seems happy with the lot he has chosen or that heaven reserved for him? [20]

He then calls the object *supreme beauty* because to him it is unique and most eminent and efficacious for drawing him to itself, and for that reason does he deem it most worthy and most noble; and yet he feels the object to be dominant and superior over him, as

---

[19]   DANTE *Convivio* 4:

> If the soul takes not its perfect "stand" it is not so disposed as to receive this blessed and divine infusion; just as, if a precious stone be ill-disposed, or imperfect, it cannot receive the celestial virtue, as said that noble Guido Guinizelli...

[20]   HORACE *Satires* i. 1. 1-3:

> Qui fit, Maecenas, ut nemo, quam sibi sortem seu ratio dederit seu fors obiecerit, illa Contentus vivat?

he is rendered subject and enslaved by it. *My one sole death* he says of jealousy because just as love has no more inseparable companion than jealousy, so love has no sense of any greater enemy; just as nothing is more an enemy to iron than rust, though that rust is generated of the same iron.

CIC.   Now since you have begun by this method, proceed to show point by point what remains.

TANS.   I shall do so. Next he says of love, *It shows me paradise.* By this he means that love is not blind of itself, and renders certain lovers blind not because of its nature, but because of the ignoble dispositions of the subject as it happens that nocturnal birds become blind in the presence of the sun. With respect to itself, therefore, love illumines, makes clear, opens the intellect, makes all things penetrate and spurs miraculous impulses toward the good.

TANS.   I'm quite certain the Nolan shows this in another one of his sonnets:

Love who shows me so high a truth that
it opens black portals of diamond, enters
its deity through the eyes and by the sight
is born, lives, is nourished and reigns eternally [21]

and makes me perceive how much heaven,
earth and hell conceal. Love brings to light
the true forms of absent things, regains force
and with a sure dart stabs and ever wounds the
heart, uncovers what is within.

Oh, therefore, vile herd, heed the truth,
lend your ear to my words that are not fallacious,
senseless and squint-eyed ones, open, open your
eyes, if you can.

---

[21] DANTE *Vita nuova* 20:

Beltade appare in saggia donna pui,
Che piace agli occhi sì che dentro al core
Nasce un desio della cosa piacente...

Beauty appears in wise lady then, which is so
pleasing to the eyes, that within the heart a
desire is born for the pleasing thing...

You believe the boy, because you understand
little; because you change swiftly, to you he
seems fleeting; in your blindness, you call him blind.

Love therefore shows him paradise because it makes
him know, understand and accomplish the highest
things, or because it gives grandeur at least in appearance to the things loved. *Fate snatches paradise away*
he says, for often fate does not concede to the deceived
lover all love has shown him, inasmuch as what he
sees and longs for is distant and opposed to him. *It
presents me with every good,* he says of the object, because the thing which love points out to him seems
to him unique, principal and ultimate. *It withdraws
it from me,* he says of Jealousy, not because it
actually wrings *every good* from his presence and from
his view, but because it makes the good no longer a
good but an agonizing evil; the sweet no longer sweet
but an agonizing languor. Therefore *the heart,* that
is to say, the will finds joy, and finds it in that very
will through the power of love regardless of the out-
come. *The mind,* in that part that recognizes that it
partakes of an ungracious fate has grief. *The spirit,*
otherwise called the natural affection, *finds refresh-
ment* in being captivated by that object which gives
joy to the heart and can satisfy the intellect. *The soul*
as the passive and sensitive substance *has a heavy
burden* because it finds itself oppressed by the heavy
weight of the jealousy which torments it.

After a consideration of his state, he adds a woeful
lament, and says, *Who will rescue me from the conflict* and give me peace; who will separate *that which
wearies me* and condemns me from that which pleases
me, and open heaven's gates to me, so that the burning
flames of my heart may be sweet and my tears be
happy? Then, continuing his proposal, he adds:

Oh, Destiny, my enemy, go torment others.
And you, Jealousy, go forth from the world. That
noble visage and insatiable Love alone, assisted
by their royal attendants can accomplish everything;

for love snatches me from life, she from
death, she gives me wings, he burns my heart;
he kills my soul; she revives it; she is my
sustainer and he is my bereaved burden.

But what have I to say of Love, if Love and
her noble visage are only one being or one form, [22]
if by the same command and law

they leave one imprint in the center of my
heart? They are not two then. They are one
which make my lot joyous and melancholy.

Four principles and extremes of two contraries he
would reduce to two principles and one contrariety.
This is why he says, *Ah me, torment the others,* which
is to say, it is enough, *oh my destiny,* that you have
oppressed me to this extent, and (since you cannot
exist without activity) turn your fury elsewhere. And
you, Jealousy, *go forth from the world,* because one of
the other two which remain will be able to take your
vicissitudes and functions upon itself: for you, my destiny, are not other than my Love, and you, Jealousy,
are not foreign to Love's substance. Therefore it is Love
that remains to deprive me of life, to burn me, to give
me death and to put all its weight upon my bones. As
for her noble visage, it remains there to snatch me
from death, to give me wings, to revive and sustain me.
Finally, these two principles and one contrariety he
reduces to a single principle and to a single efficacy,
when he says: *but what have I to say of Love?* If her
visage belongs to his empire, which is none other than
that of Love; if then the law of Love is the same as her
law; if the impression of Love sealed in my heart is
certainly none other than her impression, what need is
there, then, having called it a *noble visage,* to speak of
it again as an *insatiable Love?*

END OF THE FIRST DIALOGUE

---

[22] See above, p. 86 and note 7.

TANS.   Here the frenzied one begins to reveal his passions and disclose the wounds which are represented as wounds of the body, but are substantially or essentially wounds of the soul; and he speaks thus:

I who carry the lofty banner of love, have frozen hopes and burning desires: at one and the same time I tremble, freeze, burn and sparkle, I am dumb, and I fill the sky with ardent shrieks. [1]

My heart throws off sparks, while my eyes distil water; and I live and die, laugh and lament; the waters remain living, and the fire does not die, because I have Thetis in my eyes and Vulcan in my [heart. [2]

I love another and despise myself; [3] but if I spread my wings, the other is changed to

---

[1] SANNAZARO *Rime* 29:

Udrite il pianto, e i miei gravi lamenti;
Udrite, selve, i dolorosi accenti...

Hear my cries and my grave lamentations;
oh forests, hear my dolorous accents...

[2] Bruno cites these two quatrains in *De vinculis in genere*, in Opera, III, 9, 658-659, in which he employs an analogous theme of the contrary passions of the lover: "quem Cupidinis vincula invaserint, uno eodemque igne atque laquei sensu videbitur cogi ad exclamandum et tacendum, laetitiam tristitiam...".

[3] PETRARCH *Rime* 134:

Pace non trovo, e non ho da far guerra;
E temo e spero, ed ardo e son un ghiaccio;

stone; the other is raised to heaven, if I
am thrust below;

the other always flees, if I ceaselessly
pursue; if I call, there is no reply, and the
more I seek, the more is hidden from me. [4]

A propos of this poem I would like to return to
what I was saying a little while ago. It is not necessary to tire one's self out proving what is so evident:

E volo sopra 'l cielo, e giaccio in terra,
... ...
Veggio senza occhi; e non ho lingua e grido;
E bramo di perir, e cheggio aita;
Ed ho in odio me stesso ed amo altrui...

I have no peace, yet all my war is ended;
I fear and hope, and burn, yet I freeze like ice.
I fly up to the sky, and yet lie prostrate upon the earth.
... ...
I see, yet have no eyes; and I complain without
speech; I long to perish, yet I ask for help; I
love another and thus hate myself...

PIERRE RONSARD, Les Amours, ed Paul Laumonier (Paris, 1957), IV, 16:

J'espere et crains, je me tais et supplie,
Or je suis glace, et ores un feu chault et
... ...
Rien ne me plaist, si non ce qui m'ennuye,
... ...
J'ay l'espoir bas, j'ay le courage hault,
... ...
Cent foys je meur, cent foys je prens naissance.

See also OVID Ars amoris (Everyman), i, p. 44:

Beneath the shade of Venus' marble fane:
Here he that love consults, him love affects
Who loves another and himself neglects;
Here in this place the eloquent are dumb—
Who plead for others, for themselves are come...

[4] ANTONIO EPICURO, La Cecaria (Venezia, 1533), p. 3:

Che del tuo mal si ride, che ti fugge,
Che t'arde, ti distrugge e si nasconde,
Che mai non ti risponde...

For she laughs at your plight, for she flees
from you, she who burns you, destroys you and
runs away, and never answers your cry...

nothing is pure and unmixed (and, as some used to say, nothing that is a composite is a true entity; for composite gold is not pure gold and mixed wine is not true and pure wine); moreover, all things are made of contraries, and because of this composition in all things never do the affections which engage us bring us delight without also bringing something bitter. In fact, I shall go further; if it were not for the bitter in things there would not be delight, just as hard labor makes us find delight in rest; separation is the cause of our finding pleasure in union; and if we investigate the matter generally, it will always be found that one contrary is the occasion for the other contrary's desirability and pleasure.

CIC.   Then there is no delight without its contrary?

TANS.   Definitely not, just as without its opposite there is no pain, as the Pythagorean poet expresses it when he says:

> Hinc metuunt cupiuntque, dolent gaudentque, nec auras
> Respiciunt, clausae tenebris et carcere caeco. [5]

Such are the consequences of the composition of things. This is how it happens that none is satisfied with his lot, except some insensate and stupid person, satisfied so much the more as he finds himself in the last degree of the obscure phase of his folly; for then he has little or no apprehension of his evil, he enjoys the present without fear of the future, he is fully content with himself and with the world which surrounds him, and he has no remorse or care for what is or may be; and finally, he has no sense of the contrariety represented by the tree of the knowledge of good and evil.

CIC.   From this we see that ignorance is the mother of felicity and sensuous happiness; and this same happiness is the garden of paradise of the animals, as it is made clear in the dialogues of the *Cabala of the*

---

[5] VIRGIL *Aeneid* vi. 733-734:

> ...they fear and desire, sorrow and rejoice;
> nor do their eyes pierce the air while barred
> in the blind darkness of their prison house...

Pegasian Horse and in that which the wise Solomon says: *Who increases wisdom, increases sorrow.* [6]

TANS. From this we learn that heroic love is a torment, because it does not rejoice in the present, as animal love does, but in the future and the absent; and its contrary awakens in it ambition, emulation, suspicion and fear. Thus one of our neighbors said one evening after dinner: Never was I so happy as I am now; —Giouanni Bruno, father of the Nolan, replied:—Neither were you ever more mad than now.—

CIC. Do you mean then, that he who is sad, is wise, and he who is sadder is even wiser?

TANS. No, in fact I mean that in these is another species of madness, and one much worse.

CIC. If he who is content is mad, and he is who is sad is mad, then who has wisdom?

TANS. He who is neither content nor sad.

CIC. Who then? He who sleeps? He who has no feeling? He who is dead?

TANS. No; but he who endures, observes and understands; who, considering the evil and the good, holding the one and the other as something variable and subject to movement, mutation and change (so that the end of one contrary is the beginning of the other, and the extreme stage of one is the commencement of the other), takes care neither to humiliate himself, nor become puffed up with pride, moderates his inclinations and tempers his desires; for him it is an established fact that pleasure is not pleasure, because he is ever aware of its limits, and in the same way pain to him is not pain, because he is aware of its limits by the power of reflection. In this manner the wise man holds all mutable things as things which do not exist, and he believes these are nothing else but vanity and nothingness, because the same proportion exists between finite time and eternity that exists between the mere point and the line.

CIC. So that never can we appropriately hold the view that we are content or discontent without also holding that we are mad and without expressly confessing it; and no one who debates the question and thus participates in it will be wise. Consequently in the end everyone will be mad.

---

[6] Ecclesiastes I, 18.

TANS.

I do not intend this conclusion; for I would call him most wise who could truly express one of his contrary states occasionally by means of the other:— Never have I been less happy than now;—or again:— Never have I been less sad than now—.

CIC.

But where two contrary feelings are evident, how is it that you do not see two contrary qualities? I mean, why do you understand the minimum happiness and the minimum sadness as two virtues and not as one vice and one virtue?

TANS.

For the reason that both contraries in excess (that is, when they begin to go beyond their limit) are vices, for they exceed their range; and inasmuch as these move toward the lesser degree they become virtue because they are contained and enclosed within their extremes.

CIC.

How is the state of lesser content and the state of lesser sadness not one virtue and one vice, but two virtues?

TANS.

I say further that they are one and the same virtue; for where there is contrariety there is vice; and contrariety is there above all where the extreme is; the greater contrariety is nearest to the extreme, and least contrary or no contrary at all is in the middle where the extremes meet and become one and indifferent. For example, between the extremes of hot and cold is the more cold, and in the middle is the point you can call either hot or cold, or neither hot nor cold, a point at which no extremes are found. In the same way he who is the least content and the least happy is at the degree of indifference, and finds himself in the house of temperance where virtue resides and the condition of a strong soul, which does not give way to the south wind or the north. [7]

This is the reason why, to come to our point, the heroic frenzy, which our present discourse somewhat clarifies, differs from other more ignoble frenzies not as virtue differs from vice, but as vice practised in a divine way by a more divine subject differs from vice practised in a bestial way by a more bestial subject. Therefore, the difference is not according to the form of vice itself, but according to the subjects who practice it in different ways.

---

[7] ARISTOTLE *Nichomachean ethics* ii. 1109a.

CIC.

From what you have said, I can very well infer the state of this frenzied lover who says, *I have frozen hopes, and burning desires*, because he is not in the temperance of indifference, but in the excess of contraries, his soul in discord; if he trembles in frigid hopes, he burns in hot desires; and if his insatiability wrings shrieks from him, fear renders him dumb; he throws off sparks from his heart for the love of another, and in compassion for himself tears flow from his eyes; he dies in the laughter of another, lives in his own complaints; and as one who no longer belongs to himself, he loves another and despises himself. Similarly physicians say that matter hates its present form in proportion to its love of the form that it does not have. And thus the eighth verse concludes with the war which the soul has within itself; and then, when the poet says in the sestet, *but if I spread my wings, the other is changed to stone*, and in what follows, he shows the suffering imposed upon him by the war he wages with the contraries external to him.

I recall having read this sentence in Iamblicus, where the Egyptian mysteries are treated, *Impius animam dissidentem habet; unde nec secum ipse convenire potest, neque cum aliis.* [8]

TANS.

Now listen to another sonnet whose import follows upon what has been said:

Ah, what a condition, what a nature, or what a destiny is mine! I endure a living death, and a dead life! Ah me! love has killed me by such a death, so that I am deprived of both life and death.

Drained of hope at the gates of hell, overflowing with desire, I reach out to heaven; and as an eternal slave to two contraries, I am banished from heaven and from hell.

---

[8] "Impiously he has a divided will; therefore he can live neither with himself nor with others." This statement Bruno erroneously attributes to Iamblicus. The probable reason for the error is that in the edition of Ficino, *Opera* (Bale, 1561), II, extracts from a commentary of Proclus and Plato's Alcibiades follow immediately a translation from Iamblicus' *De myst. Aegypt.* See GIORDANO BRUNO, *Des fureurs héroïques*, trans. Paul-Henri Michel (Paris, 1954), p. 174 and note 10.

There is no respite for my pain, [9] because between two burning wheels, one which draws me here, the other there,

like Ixion, I must pursue myself and escape myself, [10] because the spur and the bit provide a contrary lesson to my doubtful discourse.

He shows how he endures the division and discord within himself. The discord occurs when the affection, leaving the middle region and final goal of temperance, tends to one and the other extreme; and when the affection is transported high or to the right, it is also transported below and to the left.

CIC.   How does that affection which is neither exactly at one or the other extreme fail to come within the state or bounds of virtue?

TANS.   Affection is in the state of virtue when it establishes itself in the mean, departing from the one and the other extreme; when it tends to the extremes, inclining to one or the other of them, it falls short of virtue so much that it becomes a double vice; and vice consists in this, that a thing deviates from its own nature whose perfection consists in unity; and the composition of virtue is at the point where the contraries unite.

Here, then, is how he is dead though living, and alive while dying; as when he says, *I endure a living death and a dead life.* He is not dead, because he lives in the object, he is not alive, because he is dead to himself; he is deprived of death, because he nurtures thoughts in the object; he is deprived of life, because in himself he neither can vegetate nor sense anything. Besides, he is most base when he considers the loftiness of the intelligible object and realizes the weakness of his power. He is most lofty through the aspiration of the heroic desire that carries him far above the limit of his own nature, most lofty through the intellectual appetite whose operation and design is not to join his desire to its object; and he is most base because of the

---

[9]   PETRARCH *Rime* 22.
[10]   OVID *Metamorphoses* iv, 461:

There whirls Ixion on his wheel, there pursuing himself and fleeing all in one...

violence brought upon him by the contrary sensuality weighing down toward the inferno. Therefore, finding himself rising and falling, in his soul he feels the greatest discord possible, and he remains confused by the rebellion of the sensuality which spurs him to the point where reason, acting in a contrary way, restrains him. This is precisely what is shown in the following dialogue. Here reason interrogates in the name of Filenio, and the frenzied lover replies in the name of Pastore, who labors to watch over the flock of his thoughts, which he feeds in the homage and service of his nymph, that is, in the service of the affection of that object to which he has become enslaved.

FIL.	Shepherd boy!
PAST.	What do you wish?
FIL.	What are you doing?
PAST.	I suffer.
FIL.	Why?
PAST.	Because both life and death reject me.
FIL.	Who is responsible?
PAST.	Love.
FIL.	That mischievous one?
PAST.	That mischievous one.
FIL.	Where is he?
PAST.	In the center of my heart, strongly fixed.
FIL.	What does he do there?
PAST.	He stabs.
FIL.	Whom?
PAST.	Me.
FIL.	You?
PAST.	Yes.
FIL.	With what means?
PAST.	With her eyes, portals of heaven and hell.
FIL.	Do you have hope?
PAST.	I do.
FIL.	Pity?
PAST.	Pity.
FIL.	The pity of whom?
PAST.	Of her who tortures me night and day.
FIL.	Does she have it too?
PAST.	I don't know.
FIL.	You're mad.
PAST.	But what if such madness is pleasant to the soul?
FIL.	Does she promise anything?
PAST.	No.
FIL.	Does she refuse?

PAST.   Not even that.
FIL.   Is she silent?
PAST.   Yes, because decorum has taken the boldness from me.
FIL.   You're raving.
PAST.   Why?
FIL.   Because you suffer.
PAST.   I fear her disdain more than I do my torments.

He tells of his intense pain, he laments of his love certainly not because he loves (for no lover really dislikes loving), but because he loves unhappily and has submitted to the arrows which are the rays of those eyes, which, accordingly as they express disdain and refusal, or on the contrary as they express benevolence and favor, become the portals which lead to heaven, or, on the other hand, to hell. Therefore he is maintained in the hope of future and uncertain mercy, and in the condition of present and certain martyrdom. And even though his own madness may be clearly evident to him, never does he manage to correct himself of it at any point; nor can he even conceive of it as unpleasant; and the more he errs because of that madness the more he delights in it, as he shows us where he says:

> May it never be that I lament of love,
> for without love I never would be happy.

Next he shows another species of frenzy, nourished by a certain light of reason, a species which excites fear and destroys the madness already mentioned, so that it does not lead to any act that would irritate or disdain the thing loved. Therefore, he says his hope is founded upon the future, although nothing is promised or denied him; for he is silent and asks nothing for fear of offending chastity. He does not dare explain himself or make any proposal which could avail to exclude him by a rejection, or assure him by a promise; for in his mind the evil that could come to him in the one case weighs more than the good that could come to him in the other. He shows himself, then, more readily disposed to suffer his particular torment forever than to risk opening the door to what might be an occasion of trouble and sadness to his beloved object.

CIC.   This proves his love is truly heroic, for he wishes for himself the favor of her spirit and the good will of affection as objects more important than her corporeal beauty, a beauty in which the love he has for the divine is not satisfied.

TANS. You know very well that there are three species of Platonic raptures. One tends to the contemplative or the speculative life; one toward the active or moral life and the last toward the life of idleness and voluptuousness; similarly there are three species of love: one which from the aspect of the corporeal form rises to a consideration of the spiritual and the divine; another which perserveres only in the delight of the sight and in conversation; and finally another which descends from the sight to the concupiscence of the touch. [11] Of these three modes others are composed, accordingly as the first is accompanied by the second or by the third, or as all three concur together; and beyond this each one of these is multiplied into others besides, according to the affections of the frenzied lovers which tend either more to the spiritual or more toward the corporeal object or toward both of them equally. As a result, among those who are found in this band, imprisoned as they all are in love's snare, some propose for the accomplishment of their desire to gather the fruit of the tree of corporeal beauty, and, failing in this satisfaction (or at least in some hope of it), they deem derisive and vain every other amorous labor. This is the way of those who are of a barbarous mind, who neither can nor desire to attain greater dignity for themselves by loving worthy things, by aspiring toward illustrious things, and higher still, by applying their ardors and their deeds to divine things; for to such ardors and deeds nothing but heroic love can more generously and efficaciously supply the wings. The goal others propose for themselves is the fruit of gratification they take from the aspect of beauty and grace of spirit which shines and radiates in bodily charm; and although some of these love the body and long very much for union with the body, lament its inaccessibility and are saddened by separation from it, they always fear their claim to it might deprive them of the affability, conversation, friendship and concord most important to them; for the assurance of the success of their efforts could not be greater than the fear of losing

---

11 PLATO *Phaedrus* 252-256; see also MARSILIO FICINO, *Commentarium in convivium* (1484), Oration VI, viii, pp. 149-150.

the favor they look upon as a thing so glorious and worthy.

CIC.   Because of the many virtues and perfections found in the human mind, Tansillo, it is worthy to seek, accept, nourish and preserve such a love; but one must still take great care not to debase himself by becoming obligated to an unworthy and degraded object, lest he participate in its ignobility and indignity. I believe this was the significance of the counsel given by the poet of Ferarra:

> Seek to rescue him who steps into love's
> snare without having your wings entangled. [12]

TANS.   To tell the truth, an object of no greater splendor than beauty of the body is not worthy of being loved for any other purpose than to propagate the species (as they say); and it seems to me proper to the swine and the horse to be tormented for that purpose; as for myself never have I been more fascinated by such a beauty than I am now over some statue or painting, for these, it seems to me, are things of the same order. It would be then a great shame for a noble spirit to say, speaking of a filthy, vile, sluggish and ignoble soul (no matter how excellent its corporeal dress), *I fear her scorn more than my torment.*

END OF THE SECOND DIALOGUE

---

[12] LUDOVICO ARIOSTO *Orlando Furioso* 24, 1:

> Chi mette il piè su l'amorosa pania,
> cerchi ritrarlo, e non v'inveschi l'ali...

Third Dialogue

TANS.   There are many species of frenzies and these may be all reduced to two sorts. The first accordingly displays only blindness, stupidity and an irrational impulse which tends to bestial folly; the second consists in a certain divine rapture [1] which makes some become superior to ordinary men. The frenzies of the last sort are divided into two species; for some of those who experience them, because they have become habitations of the gods or divine spirits, speak and do admirable things for which neither they themselves nor anyone else understand the reason; [2] and these commonly have been raised to this state from having first been undisciplined and ignorant and void of any spirit and sense of their own; in them, as in a room which has been scoured, is introduced a divine sense and spirit which has less chance of revealing itself in those who are endowed with their own sense and reason, for sometimes it is necessary that the world devoutly believe that it is given to some men to speak and act under the influence of a superior intelligence, inasmuch as their speech does not arise from their own study and experience; consequently, the multitudes may justly show greater admiration and faith in men so endowed. Others, because of a custom or habit of contemplation, and because they are naturally endowed with a lucid and intellectual spirit, when under the impact of an internal stimulus and spontaneous fervor spurred on by the love of divinity, justice,

---

[1] For Bruno's use of *abstrazione* in the Latin sense of *raptus*, see GIORDANO BRUNO, *Des fureurs héroïques*, ed. Paul H. Michel (Paris, 1954), p. 200.

[2] PLATO *Ion* 542.

truth and glory,[3] by the fire of desire and inspired purpose, they make keen their senses and in the sulphurous cognitive faculty enkindle a rational flame which raises their vision beyond the ordinary. And these do not go about speaking and acting as mere receptacles and instruments, but as chief inventors and authors.

CIC.   Which of these two species do you deem the superior?

TANS.   Those who are of the first sort have within them a great dignity, power and efficacy inasmuch as they harbor the divinity. But those who belong to the second class are of their very selves more worthy, powerful and efficacious; they are divine. Those who belong to the first are worthy in the same way as the ass who carries the sacraments;[4] those who belong to the second have a worthiness that is truly sacred. In those of the first class the divinity is considered and viewed according to its effect and is admired, adored and obeyed; in those of the second, the excellence of their special humanity is considered and brought to light.

Now we come to our purpose. These frenzies of which we speak, and whose manifestations are seen in these dialogues, do not arise from forgetfulness, but from a remembrance.[5] They are not undirected frenzies, but love and desire for the beautiful and the good, a model of perfection one proposes to attain for himself by being transformed into its likeness.[6] It is not the rapture of one caught in the snare of bestial passion under the law of an unworthy fate; but a rational force following the intellectual perception of the good and the beautiful comprehensible to man to whom they give pleasure when he conforms himself to them, so that he is enkindled by their dignity and light, and is invested with the quality and condition which make him illustrious and worthy. By intellectual contact with that godlike object he becomes a god;[7] and he has thoughts of nothing but

---

[3] PLATO Symposium 210.
[4] BRUNO, Cabala, in Opere italiane, II, 235 and note 4.
[5] PLATO Phaedrus 250.
[6] Ibid., 255; see also Symposium 211.
[7] PLATO Phaedrus 252-253.

things divine and shows himself insensible and impassible to those things which ordinary men feel the most and by which they are most tormented; he fears nothing, and in his love of divinity he scorns other pleasures and does not give any thought to his life. It is not the melancholy frenzy which—beyond counsel, reason and prudence—will make him stray at the mercy of chance and carry him in the flow of its ruinous tempest, as those who, having transgressed certain laws of the divine Adrastia, [8] were condemned to the butchery of the Furies and to the loss of all peace by a conflict that was physical, arising from seditions, ruin and maladies, as well as spiritual, arising from the loss of harmony between the rational and appetitive powers; but it is a heat enkindled in the soul by the sun of the intellect, and a divine force which sets wings upon him; [9] so that always bringing him closer to the intellectual sun, rejecting the rust of earthly cares he becomes gold proven and pure, acquires the feeling of divine and internal harmony, and conforms his thoughts and acts to the common measure of the law innate in all things. [10] He is not as one inebriated by the vessel of Circe who goes from ditch to ditch and from rock to rock, plunging and stumbling; nor is he like a variable Proteus always changing himself from one appearance to another, without ever finding any place, or mode, or manner of settling or fixing himself, but without disturbing his balance he conquers and overcomes the terrible monsters; and if he happens to decline, he returns easily to the sixth sphere, [11] thanks to those profound instincts within him which are like the nine Muses who dance and sing around the splendor of the universal Apollo; and beneath sensible images and material objects he perceives the laws of divine wisdom. It is true that sometimes, having for an escort Love, who is twofold, and because he sees himself often defrauded of the fruits of his efforts by some rising

---

[8] Adrastia in Plato and Plotinus is a symbol of wisdom and justice. See PLOTINUS *Enneads* 3, 2, 13; also PLATO *Phaedrus* 248 C.

[9] PLATO *Phaedrus* 246-247, 248 C.

[10] PLOTINUS *Enneads* 4. 7. 10.

[11] The sixth planetary sphere represents intelligence and order; see BRUNO, *Des fureurs héroïques,* p. 50.

obstacle, then, like one insensible and frenzied, he overthrows the love of what he cannot understand; and thus confused by the abyss of divinity, sometimes he gives up the contest. Then he returns, nevertheless, and forces himself to attain by his will what he cannot attain by his reason. It is also true that he usually wanders at random and transports himself now toward one and now toward another form of twofold Eros, for the chief lesson love teaches him is to contemplate the shadow of the divine beauty (when he cannot contemplate its direct reflection), as, for example, the suitors of Penelope amused themselves with her servants when they were not permitted to converse directly with the mistress herself. Now to conclude, you can understand from what has been said, of what species this frenzied one is, whose image is shown us in these verses:

If the butterfly wings its way to the sweet light that attracts it, it is because it knows not that the fire is capable of consuming it; [12] if the thirsty stag runs to the brook, it is because he is not aware of the cruel bow. [13]

If the unicorn runs to its chaste nest, it is because he does not see the noose which is prepared for him. In the light, at the fount, in the bosom of my love's light, I see the flames, the arrows and the chains.

If my languishing is so sweet to me, it is because the heavenly face delights me so, and because the heavenly bow so sweetly wounds;

And because in that knot is bound up my desire, I suffer eternally through the fire of my heart, the arrow in my breast, and the yoke upon my soul.

Here he shows that his love is not like that of the butterfly, the stag or the unicorn, who would run away if they had some idea of the fire, of the arrow

---

[12]   TANSILLO, p. 155, pp. 162-163; see also PETRARCH *Rime* 141.
[13]   PIERRE DE RONSARD, *Les Amours*, IV, p. 52.

and the noose, and who perceive nothing but what pleases them. He, on the contrary is guided by a most keenly felt and only too lucid frenzy, which makes him love that fire more than any other consideration, that wound more than any state of health, those chains more than any other freedom. For this evil is not an evil absolute; it is an absolute evil only with respect to what is held good according to a certain opinion. And this opinion is as fallacious as the condiment old Saturn used (for his dinner), [14] when he devoured his own sons. For this evil in the eyes of the absolute and of eternity is understood either as a good, or as a guide leading us to the good; for this fire is the burning desire for divine things, this arrow is the impact of the ray of the beauty of the divine light, these yokes are the species of the true and the good which unite and join our minds to the primal truth and the supreme good. I spoke in this sense when I said:

By so beautiful a fire and so noble a yoke,
beauty enkindles me, and chastity entangles me,
so that I must be happy in fire and in slavery;
liberty I must flee and I must dread the ice.

The conflagration is such that I burn yet
am not consumed, and the yoke is such that the
world celebrates it with me; neither am I frozen
by dread, nor undone by grief; but my ardor is
tranquil, my burden sweet.

I perceive so lofty a light that I am enkindled by it, and a noose devised of such rich
yarn, that as contemplation grows, desire dies.

Because so beautiful a flame enkindles my
heart, and the desire for so sweet a bond compels
me, darkness is my servant and my ashes glow. [15]

All loves (if they are heroic, and not purely animal, the physical means by which those enslaved by
nature are called to procreation) have divinity for
their object and tend to the divine beauty, a beauty

---

[14] Parentheses mine.
[15] TANSILLO, p. 14.

which first communicates itself to the souls and is resplendent in them, and then, from the souls, or better still, through the souls, is communicated to the body. Thus a well-ordered passion loves the body, or corporeal beauty, only because it is a sign of the beauty of spirit. In fact we become enamoured of the body because of a certain spirituality we see in it, a spirituality called beauty, and a beauty which does not consist in larger or smaller dimensions, in determined colors or forms, but in a certain harmony and concordance of the bodily members and hues. To the most acute and penetrating senses, this harmony of members shows a certain sensible affinity to the spirit; consequently, those who are so endowed fall in love more easily and more intensely and they also fall out of love more easily and are more intensely provoked. This ease and intensity can be explained by a change that takes place in the beloved object as it expresses an ugly spirit made evident in some gesture or in some expressed intention; so that as such ugliness passes from the soul to the body, the body no longer seems beautiful as it once seemed. The beauty of the body, then, has the power to enflame, but certainly does not have the power to bind the lover and keep him from fleeing from it, if that body is not assisted by the grace of spirit he desires or by chastity, courtesy and sagacity.

CIC.   Do not believe that this is always so, Tansillo; for sometimes, although we discover a vicious spirit, we remain none the less enflamed and ensnared by it; for although the reason recognizes the evil and baseness of such love, it does not have the virtue of throwing off the disordered appetite. I believe the Nolan found himself in a like disposition when he wrote:

Ah me, a frenzy constrains me to cling to my evil; which makes love appear to me as a supreme good.

Ah me, my soul is not troubled that it is always bound by contrary counsels; with that cruel tyranny which nourishes me in torment and has had power to exile me from myself, I am content more than with my freedom.

I hoist my sails to the wind, which pulls
me toward the odious good and leads me to sweet
tempestuous damnation. [16]

TANS.   This occurs when both souls are vicious and as
though spotted by the same ink, so that, because of
their likeness love is aroused, enkindled and confirmed.
Thus the vicious meet each other in the practice of
the same vice. And here I shall not be silent about
what I know from experience. I have had occasion
to discover in a certain soul vices particularly abhor-
rent to me such as sordid avarice, a most gravelling
appetite for gain, ungrateful disregard of favors and
courtesies granted, and an affinity for certain thor-
oughly vile persons (the most displeasing of all vices,
because it leaves the lover with no hope of ever being
or becoming more worthy of his beloved, or of becoming more acceptable to her); none the less I did not
fail to burn for her corporeal beauty. But the reason?
I loved her without good will, and if this had not
been the case, I would have been made sad rather
than happy by her shamefulness and wretchedness.

CIC.   That distinction between loving and having good
will toward the beloved is very apt and to the point.

TANS.   Yes. For toward many do we have good will,
which is to say, that we wish them to be wise and
just, but we do not love them, because they are ini-
quitous and ignorant. And many we love because
they are beautiful, but we do not wish them well
because they do not merit it; and among those things
he deems his beloved does not merit, the first is the
love he has for her. For that reason he regrets loving
her the more he is unable to refrain from doing so.
This is the regret he refers to when he says, *Ah me, a
frenzy constrains me to cling to my evil.* But he was
in an opposite frame of mind when he said, either referring to another corporeal object in similitude, or to
a truly divine subject:

Though you inflict upon me such cruel tortures, even so I thank you, and owe you much,

---

[16]   DANTE, *Inferno*, 5. 28-48.

Love, for you opened my breast with so generous
a wound and have so mastered my heart, [17]

that it truly adores a divine and living
object, most beautiful image of God on earth. [18]
Let him who will, think my fate cruel because
it kills in hope and revives in desire.

I am nourished by my high enterprise; and
although the soul does not attain the end desired and is consumed by so much zeal,

it is enough that it burns in so noble a
fire; it is enough that I have been raised to
the sky and delivered from the ignoble number.

Here his love is completely heroic and divine. And
I would understand it as heroic and divine, even
though because of it he speaks of himself as afflicted
by such cruel tortures; for every lover who is separated from the beloved (to which, joined by his desire, he
would also be joined in act) finds himself in anguish
and pain, crucifies himself and torments himself. He
is so tormented, not only because he loves and is
conscious that his love is most worthily and nobly
employed, but because his love is deprived of that
fruition which it would attain if it had arrived at the
end toward which it tends. He does not suffer because
of that desire which enlivens him, but because of the
difficulty of the labor which martyrs him. Thus others
consider him as being in an unhappy condition because of the fate which seems to have condemned
him to these torments; as for himself, despite these
torments, he will not fail to recognize his debt to Love
and will not fail to render thanks to it, because it has

---

[17]   ANTONIO EPICURO, *Cecaria*, p. 40.

[18]   DANTE *Vita nuova* 14:

For love... exultant grows and takes such assurance that he
smites among my afflicted senses... so that he alone remains
to gaze upon you.

Ch'Amor... prende baldanza e tante sicurtate,
Che fiere tra' miei spirti paurosi...
Sicch' ei solo rimane a veder vui.

brought an intelligible form before his mind. For in that intelligible form, although he is enclosed within the prison of the flesh during this earthly life, bound by his sinews and confined by his very bones, he has been permitted to contemplate an image of the divinity more exalted than would have been possible had some other species and similitude of it been offered him.

CIC. The god-like and *living object* of which he speaks, then, is the highest intelligible aspect of the divinity he is able to experience for himself; and it is not some corporeal beauty which would obscure his thought as it appears superficially to the sense.

TANS. True, because no sensible thing or species of it can be elevated to so much dignity.

CIC. Then how is it that he mentions the intelligible form as the object (of his love) if, as it seems to me, the true object is the divinity itself?

TANS. The divinity is the final object, the ultimate and the most perfect object, but it certainly cannot be found here below where we can see God only as in a shadow or a mirror; and for that reason the divinity can be the object only in similitude, and not a similitude abstracted and acquired from corporeal beauty and excellence by virtue of the senses, but a similitude the mind can discern by virtue of the intellect. When it has reached this state, the mind begins to lose love and affection for every other sensible as well as intelligible object, for joined to that light it becomes that light, and consequently becomes a god. For the mind draws the divinity unto itself, being in God by the effort to penetrate the divinity (as much as it can); and God is in that mind, for after having penetrated the divinity the mind will conceive the divinity and (as much as it can) will receive the divinity and retain a concept of it. Now the human intellect feeds itself upon species and similitudes in this inferior world, inasmuch as it is not permitted to contemplate the beauty of the divinity with purer eyes. Thus he who arrives at some most excellent and most beautifully adorned edifice and considers it in each detail, is pleased, contented and filled with a noble wonder; but then should it happen that he also see the lord of these images in his incomparably greater beauty, he would abandon every concern and thought of such images, turn and become completely intent upon the contemplation of that lord. Such is the difference between the state in which we see the divine beauty in its intelligible aspects which are drawn from the divine beauty's effects, operations, designs, shadows and similitudes, and that other state in which we might be permitted to see it in its own unique being.

Then he says, *I am nourished by my high enterprise* because (as the Pythagoreans knew) in this way the soul is turned and moves toward God, as the body moves toward the soul.

CIC.   The body, then, is not the abode of the soul?

TANS.   No; for the soul is not in the body locally, but is in it intrinsically as its form, and extrinsically as creator of its form, similar to that which forms the members and shapes the composite from within and from without. It is the body, then, that is in the soul; the soul is in the mind, and the mind either is God or is in God, as Plotinus said. [19] And just as by its essence the mind is in God who is its life, similarly by its intellectual operation and the consequent operation of the will, the mind refers itself to its own light and its beatific object. It is therefore with dignity that this passion of the heroic frenzy feeds itself upon so high an emprise. Although the beatific object is infinite, and in act perfectly simple, and although our intellective potency is unable to comprehend the infinite, except in speech or in a certain manner of speaking, or, as otherwise said, by a certain potential reason and natural disposition, he of whom we speak does not differ from one who would aspire toward the immeasurable as an end where in fact there is no end.

CIC.   And this is most nobly as it should be; for, in fact, the last end ought not to have an end, otherwise it would not be the last. Therefore it is infinite in purpose, in perfection, in essence and in every manner possible.

TANS.   You speak the truth. Now in this life the peculiarity of such nourishment is that it enflames the desire more than it can satisfy it, as that divine poet shows us well in the words, *My soul languishes in the desire for the living God;* [20] and elsewhere when he

---

[19] PLOTINUS *Enneads* 4. 3. 24; see also MARSILIO FICINO *In Plotini Enneades* iv, 3. 20, 24, for Bruno's direct source.

[20] Pss. 41: 2, 3.

says, *Attenuati sunt oculi mei suspicientes in excelsum.* [21] This is why our own poet says, *And though the soul does not attain the end desired and is consumed in so much zeal, it is enough that it burns in so noble a fire.* He means the soul is consoled in this ardor and receives all of the glory possible to it in its present state, and participates in that ultimate frenzy of man, inasmuch as he is a man in the state in which he finds himself presently as we see him.

CIC.   I imagine the Peripatetics (as Averroës explained) have this in mind, when they say the ultimate happiness of man consists in attaining perfection in the speculative sciences. [22]

TANS.   It is true, and they put it very well. For in this condition of ours we cannot desire or attain greater perfection than that which is ours when our intellect through the medium of some noble intelligible species is united either to the separate substances, as some say, or to the divine mind, if we employ the idiom of the Platonists. And I shall omit discussion about the soul, or man in another state and mode of existence in which he may find or believe himself.

CIC.   But what perfection and satisfaction can man find in a cognition which is not perfect?

TANS.   Cognition can never be perfect to the extent that it shall be able to understand the highest object; but only to the extent that our intellect has the power to understand this object. It suffices that in this state of ours and in any other our intellect may perceive the divine beauty to the degree that it extends the horizon of its vision.

CIC.   But all men cannot reach that point, but only one or two.

TANS.   It is enough that all attempt the journey. It is enough that each one do whatever he can; for a heroic mind will prefer falling or missing the mark nobly in a lofty enterprise, whereby he manifests the dignity of his mind, to obtaining perfection in things less noble, if not base.

---

[21] *Isaiah* 38: 14: "My eyes are diminished as they gaze into the heavens..."

[22] Averroës, *In libros physicorum Arist. Proemium, in Aristotelis,* in *Opera omnia* (Venetiis, 1574), IV, fol. 1, H.

CIC.   Certainly a worthy and heroic death is preferable to an unworthy and vile triumph.

TANS.   A similar thought inspires the following sonnet:

Since I have spread my wings toward sweet delight, the more do I feel the air beneath my feet, the more I spread proud pinions to the wind, and contemn the world, and further my way toward heaven.

Nor does the cruel fate of Daedalus's son burden me, on the contrary I follow his way the more: that I shall fall dead upon the earth I am well aware; but what life compares with this death?

I hear the voice of my heart upon the wind: Where do you take me, adventurous one? Resign yourself, for too much temerity is rarely without danger.

I reply: fear not noble destruction, burst boldly through the clouds, and die content, if heaven destines us to so illustrious a death. [23]

---

[23] TANSILLO, p. 14; see also IACOPO SANNAZARO, *Rime, Scrittori Italiani;* Carli-Sainati (Florence, 1950) I, pp. 870-871:

Icaro cadde qui: queste onde il sanno
... ...
Avventuroso e ben gradito affanno,
   Poi che, morendo, eterna fama ottenne:
... ...

Ben può di sua ruina esser contento,
   S'al ciel volando a guisa di colomba,
   Per troppa ardir fu esaminato, e spento;
Ed or, del nome suo, tutto rimbomba
   Un mare, sí spazioso, un elemento:
   Chi ebbe al mondo mai sí larga tomba?

Icarus fell here as these waters witness
... ...
daring and most ingratiating disaster, for by
   dying he attained eternal life.
... ...

Such ruin can well content him, for by flying
   to heaven in the guise of a dove he was tried and

CIC.   I understand when he says, *It is enough that I have been raised to the sky;* but not when he says, *and delivered from the ignoble number;* unless he means that he has come out of the Platonic cavern, [24] removed from the condition of the stupid and most vile multitudes; for it is understood that those who profit from this contemplation can be only a very small number.

TANS.   You have understood it very well. Moreover, by the ignoble sod it is possible that he means the body and the sensual cognition from which he who would become united to a nature of a contrary kind must raise and disengage himself.

CIC.   The Platonists speak of two kinds of knots with which the soul is tied to the body. One is a certain vivifying act which like a ray descends from the soul to the body; the other is a certain vital quality in the body which results from this act. [25] Now in what manner do you understand that this most noble moving number called the soul [26] is disengaged from that ignoble number which is the body?

TANS.   It certainly was not meant that the soul can detach itself from the body in some physical way, but in a way peculiar to its potencies, which, not enclosed and enslaved within the bosom of matter, are sometimes as though lulled and inebriated and find themselves nevertheless occupied in the formation of matter and in the vivification of the body. Sometimes these potencies, as though reawakened and remembering themselves, recovering consciousness of their principle and origin, turn themselves to superior things and force themselves toward the intelligible world as to their native home; but sometimes the potencies tumble from the intelligible world by a conversion to inferior things beneath the fate and necessities of

---

             consumed by too much fervor;
And now so vast an element as the entire sea
             resounds with his name:
             who in this world has ever had so large a tomb?

[24]   PLATO *Republic* vii, 514; see also PLOTINUS *Enneads* 4. 8. i.

[25]   PLOTINUS *Enneads* 4. 3. 9. For Bruno's immediate source, as always, see FICINO, *In Plotini Enneades,* iv, 4, 18-19.

[26]   See ARISTOTLE *De anima* i, 4, 408b, for his opinion that the soul does not move of itself.

generation. These two drives are represented by the two kinds of metamorphoses which the present sonnet describes:

> That god who wields the resounding thunderbolt Asteria saw as a furtive eagle, Mnemosyne saw as a shepherd, Danae saw as gold, Alcmena saw as a fish, and Antiope as a satyr;

> to the sisters of Cadmus he was a white bull, to Leda he was a swan, and a dragon to the daughter of Demeter. I, because of the loftiness of my object, from the most vile subject become a god. [27]

> Saturn was a horse, Neptune a dolphin, Ibis took the form of a heifer, and Mercury became a shepherd,

> Bacchus a grape, Apollo a raven; and I by the mercy of love, am changed from a base thing into a deity.

There is in nature a revolution and a circle in virtue of which, for the perfection and aid of others, superior things incline toward the inferior, and for their own excellence and felicity inferior things are raised to the superior. But the Pythagoreans and the Platonists hold that souls, not only by a spontaneous will which brings them to an understanding of natures, but also by the necessity of an inward law written and recorded by a fatal decree, at certain times set out to seek their own destinies justly determined. And these say that if souls separate themselves from the divinity, it is not so much from a rebellious will of their own, as from a certain order in virtue of which they become inclined toward the material. Therefore, not from a voluntary intention, but from a certain mysterious consequence, they begin to fall. And this is why their tendency leads them toward the lesser good called generation. (I will use the word lesser insofar as

---

[27] OVID *Metamorphoses* vi, 106-115.

it pertains to a particular nature; but not at all as it pertains to universal nature, where nothing happens without the highest purpose which disposes of all things according to justice.) [28] Once they have occupied themselves with generation, the souls (by a new conversion which follows in turn) return once again to their superior states.

CIC.   Would those have it, then, that the souls are impelled by the necessity of fate, and that they have no counsel of their own to guide them at all?

TANS.   Necessity, fate, nature, counsel, will, in things justly and impeccably ordered, all concur. Besides, according to the inference of Plotinus, [29] some would have it that certain souls can escape their peculiar evil, those souls which, before they are confirmed in their corporeal garb, recognizing the danger, take refuge in the mind. Because the mind raises them to sublime things, as imagination debases them to inferior things; the mind maintains them in rest and identity as the imagination in movement and diversity; the mind forever understands the one, as the imagination forever goes about inventing varied images. In the middle is the rational faculty which is composed of everything, as that in which concurs the one and the many, the same with the diverse, motion with position, the inferior with the superior.

Now this conversion and change is symbolized in the wheel metamorphoses, in which a man is placed at the top, a beast lies at the bottom, one half-man and half-beast descends from the left, and one half man and half beast ascends from the right. This transformation is shown in which Jove, according to the diversity of the affections and their manifestations toward inferior things, invests himself in varying appearances, which assume the forms of beasts; and the other deities likewise transform themselves into ignoble and alien forms. And on the other hand, because of the sense of their own dignity, they recover their own divine forms; just as the heroic lover, raising himself by his conception of the species of divine beauty and goodness upon the wings of his intellect

---

28 PLOTINUS *Enneads* 4. 3. 12.
29 *Ibid.*, 4. 4. 5, see also FICINO, *In Plotini, Enneades*, iv, 4, 5.

and intellectual will exalts himself toward the divinity, abandoning the form of more ignoble thing. And for that reason he said: *From a more vile creature I become a God, I change into a deity from a base creature.*

END OF THE THIRD DIALOGUE

FOURTH DIALOGUE

TANS.   Now is described the path taken by heroic love,
as it tends toward its proper object, the supreme
good, and the path taken by the heroic intellect as it
strives to attain its proper object, the primary or absolute truth. [1] All of the above is summarized in the
first poem which expresses the purpose to be developed
in the following five. Thus he says:

> The youthful Actaeon [2] unleashes the mastiffs
> and the greyhounds to the forests, when destiny
> directs him to the dubious and perilous path,
> near the traces of the wild beasts.

> Here among the waters he sees the most
> beautiful countenance and breast, that ever one
> mortal or divine may see, clothed in purple and
> alabaster and fine gold; and the great hunter
> becomes the prey that is hunted.

> The stag which to the densest places is wont
> to direct his lighter steps, is swiftly devoured
> by his great and numerous dogs.

> I stretch my thoughts to the sublime prey,
> and these springing back upon me, bring me death
> by their hard and cruel gnawing.

---

[1] For the view that heroic love and heroic intellect are two wings of·
the soul see FICINO, *In convivium*, vii, 14.

[2] For the myth of Acteon as a poetic theme see PETRARCH *Rime* 23,
156-160; EPICURO, *Cecaria*, p. 52. See also BRUNO, *Des fureurs héroïques*,
trans. Paul-Henri Michel (Paris, 1954), p. 238 and note 2. For Ovid's version
of the myth of Acteon see *Metamorphoses* iii, 173-252.

Actaeon represents the intellect intent upon the capture of divine wisdom and the comprehension of the divine beauty. He *unleashes the mastiffs and the greyhounds;* of these the greyhounds are swifter and the mastiffs more powerful, for the operation of the intellect precedes the operation of the will; but the latter in turn is the more vigorous and efficacious; since divine goodness and beauty are more lovable than comprehensible to the human intellect, and besides love moves and spurs the intellect to go before it, like a lantern, *to the forests,* uncultivated and lonely, very rarely visited and explored, with the result that few men have left the traces of their steps there. The youth is of little experience and practice, as one whose life is brief and whose frenzy is unstable. *In the dubious path* refers to the uncertain and the ambiguous reason and passion which the letter Y of Pythagoras symbolizes. [3] On the right this path shows him the more thorny, uncultivated and deserted arduous path upon which he unleashes the greyhounds and mastiffs *near the traces of the wild beasts,* which are the intelligible modes of ideal concepts. These are hidden, are pursued by few men, are visited most rarely, and do not offer themselves to everyone who seeks them. *Here among the waters,* that is to say, in the mirror of similitudes, in the works in which is resplendent the efficacy of the divine goodness and splendor—these works are represented by the symbol of the superior and inferior waters over and beneath the firmament. [4] *He sees the most beautiful countenance and breast,* that is to say, he sees the power and external operation which *can be seen* in the state and act of diligent contemplation of a mortal or divine mind, by a man, or by some diety.

CIC.   If he compares divine and human comprehension and places them within the same class, I believe that he does so not with respect to the two modes of comprehension, which are very different, but with respect to the object of contemplation which is one and the same.

TANS.   That is it exactly. He says in *purple, alabaster*

---

[3] To Pythagoras the letter Y symbolizes the two paths which are offered men. See BRUNO, *Des fureurs héroïques,* p. 238 and note 4.

[4] Gen. 1: 6, 7.

and gold, meaning the purple of divine power, the gold of divine wisdom, the alabaster of divine beauty, in the contemplation of which the Pythagoreans, Chaldeans, Platonists and others attempt to rise as best they can. *The great hunter sees:* he has understood as much as he can, and he himself *becomes the prey;* that is to say, this hunter set out for prey and became himself the prey through the operation of his intellect whereby he converted the apprehended objects into himself.

CIC.   I see. For he gives shapes according to his mode to the intelligible species and proportions them to his capacity inasmuch as they are received according to the mode of him who receives them.

TANS.   And he becomes the prey by the operation of the will whose act converts him into the object.

CIC.   I understand; for love converts and transforms into the thing loved.

TANS.   You know very well that the intellect understands things intelligibly, that is, according to its own mode; and the will pursues things naturally, that is, according to the manner in which things exist in themselves. Therefore, Actaeon, who with these thoughts, his dogs, searched for goodness, wisdom, beauty and the wild beast outside himself, attained them in this way. Once he was in their presence, ravished outside of himself by so much beauty, he became the prey of his thoughts and saw himself converted into the thing he was pursuing. Then he perceived that he himself had become the coveted prey of his own dogs, his thoughts, because having already tracked down the divinity within himself it was no longer necessary to hunt for it elsewhere.

CIC.   Then it is well said that the kingdom of God is within us, and that divinity lives within us by virtue of the regenerated intellect and will. [5]

TANS.   Precisely. Actaeon becomes the prey of his own dogs, pursued by his own thoughts, turns his feet and *directs his new steps;* is renewed for a divine course— that is, with greater facility and with a more efficacious inspiration—*toward the densest places,* toward

---

[5] Cor. I. 3: 16.

Know you not that you are the temple of God
and that the Spirit of God dwelleth in you?

the deserts, toward the region of incomprehensible things: from the vulgar and common man he was, he becomes rare and heroic, rare in all he does, rare in his concepts, and he leads the extraordinary life. It is there that *his great and numerous dogs bring him death;* thus he stops living according to the world of folly, of sensuality, of blindness and of illusion, and begins to live by the intellect; he lives the life of the gods, he feeds upon ambrosia and is drunk with nectar. Now, in the form of another similitude, he describes the manner in which Actaeon arms himself for the attainment of the object, and he says:

My solitary sparrow, [6] no longer delay making your nest in that place which clouds and fills all my thought. There, above, give the full measure of your labor, your industry and art.

Find new life there and raise your lovely offspring. Now that cruel destiny has run its full course, it no longer impedes you from your enterprise, as it used to do.

Go, a more noble refuge I desire for you—and you shall have as a guide a god who by those who see nothing is called blind.

Go, and may every god of this immense creation be merciful to you; and return not to me, since you are no longer mine.

The lover's former progress symbolized by the hunter stirring his dogs here is symbolized by a winged heart; and from the cage in which it reposed in idleness and quiet it is dispatched to build its nest up on high, and to raise its little ones there—its thoughts—the time having come in which the obstacles posed by a thousand lures without and by the natural feebleness within are no longer present. He gives the heart permission, then, to attain a more noble state for itself, and turns it to a more lofty design and purpose, now that those powers of the soul which the

---

[6] For a variant of this sonnet see BRUNO, *De l'infinito,* in *Opere,* I, 285. See also Pss. 101: 8 for an analogous symbol of the "solitary sparrow."

Platonists have already represented by the two wings are more firmly developed. [7] And as a guide to the heart he designates that god whom the vulgar in their blindness call blind and mad; and that god is love who by the mercy and favor of heaven has the power to transform the heart into that other nature to which it aspires, or, after its voyage of exile, to restore it to that state from which it was banished. That is why he said, *and return not to me since you are no longer mine,* so that not unworthily I may say with that other poet:

You have left me, my heart, and light of
my eyes, you are no longer with me. [8]

Next he describes the death of the soul, called by the Cabalists death of the kiss, [9] symbolized in the Canticle of Solomon, where the beloved lady speaks these words:

Let him kiss me with the kiss of his mouth,
because by his blows too cruel a love makes me
languish; [10]

by others this death is called sleep, as in the Psalmist's words:

If I shall give sleep to my eyes,
slumber to my eyelids, I shall find in him
peaceful repose. [11]

He then speaks for the soul as languid inasmuch as it is dead in itself, and alive in its object:

Oh frenzied ones, take care of your hearts;
for mine, too much estranged from me, led away
by a harsh and pitiless hand, finds its happy
sojourn where it is smitten and dies.

---

[7] PLATO *Phaedrus* 246 D.
[8] Pss. 37: 11. See also PETRARCH *Rime* 276, v. 14.
[9] Ad. FRANK, *La Kabbala* (Paris, 1892), p. 188.
[10] Cant. 1: 1, 5:6-8.
[11] Pss. 131: 4, 5.

My thoughts call it back at every hour; and
in revolt, foolish falcon, it no longer knows that
friendly hand, from which it has flown forth
not to return.

Wild beast, who satisfies while giving pains,
you ensnare the heart, the spirit and the soul by
your spurs, your flames and your chains,

by your glances, accents and lures; and the one
who languishes and burns and does not return, who
shall heal him, who shall cool his fire and unloose
his chains?

Here the sorrowing soul, not in real discontent,
but in the passion of a certain amorous martyrdom,
speaks as though addressing its discourse to those
who are similarly impassioned. It has dismissed its
heart, as it were, against its will, for the heart directs
its course toward an impossible goal, extends itself
where it cannot reach and would embrace what it
cannot grasp; and the more the heart is estranged
from the soul, the more does it enkindle itself toward
the infinite.

CIC. Tansillo, how does it happen that the soul in this
stage of its development is happy in its own torment?
Where does that spur come from which always stimu-
lates it beyond what it possesses?

TANS. From this which I shall tell you now. Although the
intellect has arrived at the apprehension of a certain
definite intelligible form, and the will to a desire in
proportion to that apprehension, the intellect does
not stop there; for its own light impels it to think of
that which contains every genus of the intelligible
and appetitive, until it is about to apprehend the
eminence of the source of ideas, the ocean of all
truth and good. Thus it happens that whatever species
is represented to the intellect and comprehended by
the will, the intellect concludes there is another species above it, a greater and still greater one, and
consequently it is always impelled toward new motion and abstraction in a certain fashion. For it ever
realizes that everything it possesses is a limited thing
which for that reason cannot be sufficient in itself,
good in itself, or beautiful in itself, because the limited thing is not the universe and is not the absolute
entity, but is contracted to this nature, this species

or this form represented to the intellect and presented to the soul. As a result, from that beautiful which is comprehended, and therefore limited, and consequently beautiful by participation, the intellect progresses toward that which is truly beautiful without limit or circumscription whatsoever.

CIC. This procedure seems vain to me.

TANS. Not at all, in fact, because it is neither fitting nor natural that the infinite be understood, [12] or that it present itself as finite, for then it would cease to be infinite; but it is perfectly in accord with nature that the infinite, because of its being infinite, be pursued without end, in that mode of pursuit which is not physical movement, but a certain metaphysical movement. And this movement is not from the imperfect to the perfect, but it goes circling through the degrees of perfection to reach that infinite center which is neither form nor formed.

CIC. I would like to know how by circling you can arrive at the center.

TANS. This I cannot imagine.

CIC. Then why do you say it?

TANS. Because I can say it and leave it for you to consider.

CIC. If you don't mean that he who pursues the infinite is like one who, moving along the circumference, seeks the center, I don't know what you mean.

TANS. It is other than that.

CIC. Now if you don't wish to explain it, we'll not speak of it any more. But tell me, if you will, what he means when he says that his heart *is led away by a harsh and pitiless hand?*

TANS. He uses here a similitude or metaphor borrowed from common usage, which calls cruel the object that gives no fruition, or, at best partial fruition, and is more an object of desire than of possession, so that he who has partial possession of it cannot rest in full happiness, because he still desires it with an ardor which brings him to the point of a swooning, and to the point of death.

CIC. What are those thoughts which call back the heart to retard it from so noble an enterprise?

TANS. The sensitive and other natural affections which looks to the preservation of the body.

---

12 See above, introduction, p. 41 and note 34.

CIC.   What have these affections to do with the body which can in no way be of any aid or assistance to them?

TANS.   They have nothing to do with the body, but with the soul which, too intent upon a single effort or goal, becomes remiss and shows little zeal for anything else.

CIC.   Why does he call his heart *that foolish falcon?*

TANS.   Because it knows of things above.

CIC.   Usually one calls foolish those who know less than others.

TANS.   No. As a matter of fact those are called foolish whose knowledge does not conform to the common rule, whether they tend to base things, having less sense, or to higher things, having more intellect.

CIC.   I believe you are right. Now tell me further. What are *the spurs, the flames and the chains?*

TANS.   The spurs are those new pricks which stimulate and re-awaken the affection in order to render it attentive; the flames are those rays of beauty which enkindle the man who is ready to contemplate it;[13] the chains are the details and circumstances which fix the eyes of the attention and firmly unite the intellectual powers to their object.

CIC.   What are the *glances, accents and lures?*

TANS.   Glances are the persuasions whereby the object (as though it gazed at us) presents itself to us; the accents are the persuasions the object uses to inspire and inform us; the lures are the circumstances which please and attract us. So that the heart which sweetly languishes, gently burns and constantly perserveres in its enterprise, fears that its wound may heal, that its fire will go out, and its knot be untied.

CIC.   Now recite what follows.

TANS.   Lofty, profound and living thoughts of
mine, ready to flee the maternal bonds of the
afflicted soul, and disposed as archers to
aim where the lofty idea is born;

along these steep paths, heaven allows you
to encounter the cruel beast. Remember to return and recall the heart which lies concealed
in the hand of a savage goddess.

---

[13] DANTE *Convivio* 3. 8.

Arm yourselves with the love of the domestic
fires, and curb your sight so forcefully, that

these companions of my heart shall not make
you stranger to it. At least bring tidings of
its delight and joy. [14]

Here is described the natural solicitude of the
soul made attentive to generation by the friendship
it has contracted with matter. The soul dispatches its
armed thoughts which, stimulated and spurred on by
the complaint of the inferior nature, are commanded
to call back the heart. The soul instructs its thoughts
how they are to behave, for charmed and attracted by
the object as they are, they are not too easily seduced
to remain captives and companions of the heart. Therefore the soul tells them they ought to arm themselves
with the love which burns with domestic fires, that is,
the love friendly to generation to which they have an
obligation, and of which they are to be the mes-
sengers, ministers and soldiers. The soul, then, orders
its thoughts to curb their sight, to close their eyes, in
order not to gaze upon any other beauty or goodness
than the one present to them, their friend and mother.
And the soul finally concludes that, should its thoughts
not wish to be recalled for any other duty, they at
least can return to give the soul some news of the
condition and state of its heart.

CIC. Before you proceed further, I should like you to
explain what the soul means when it says to its
thoughts, *Curb your sight so forcefully?*

TANS. I will tell you. All love proceeds from the sight,
intellectual love from the eye of the mind; sensible
love from the view of the senses. Now the word sight
has two meanings. It can mean the visual potency, that
is, the power of seeing of the intellect or of the eye;
or it can also mean the visual act, the application
which the eye or the intellect makes upon the material
or intellectual object. Thus when the thoughts are
advised to curb the sight, it is not to be understood
in the first way, but in the second, because it is the
visual potency become act which begets the affection
of the appetite, whether sensitive or intellectual.

---

[14] See above, introduction, p. 37.

CIC.   This is what I desired to hear you say. Now if the visual act is the cause of the evil or of the good which proceeds from the sight, how is it that we love and desire the sight? And how does it happen that in the matter of divine things our love is greater than our understanding?

TANS.   We desire the sight because in some way we know the good of seeing, and that the act of seeing offers us beautiful things. Therefore, we desire that act because we desire beautiful things.

CIC.   We desire the beautiful and the good, but the sight is neither beautiful nor good; in fact, it is rather an instrument of comparison or light whereby we see not only the beautiful and good, but also the wicked and the ugly. It seems to me that the sight can be beautiful or good, as we can see either white or black. Therefore, if the sight (which actively perceives) is neither beautiful or good, how can it be desired?

TANS.   It is not desired for itself, but surely because of some object, inasmuch as the apprehension of an object cannot take place without it.

CIC.   What will you say if the object is neither one of sense nor of intellect? How, I ask, can the object be desired, or even seen, if there is no knowledge of it at all, if it has not occasioned any act of intellect or sense, in fact if one doubts whether it is an intelligible or sensible, incorporeal or corporeal object, or whether it is one or two or more objects, or of one or the other nature?

TANS.   To that I would say that there exists in the sense and in the intellect an appetite and impulse toward the sensible in general. This is because the intellect desires to know all of the truth, in order to grasp all that is beautiful and good in the intelligible world. The sensitive potency wishes to be informed of all that comes within the class of the sensible, and to grasp all that appears as beautiful and good to the senses. Thus we desire no less to see things we have never seen than things we have already understood and seen. But it does not follow from this that desire does not proceed from cognition, and consequently that we desire things which we do not know. On the contrary, I hold it to be well established that we do not desire what is unknown. For if things are unknown with respect to their particular natures, they are not unknown with respect to their general natures; in the visual potency one finds everything which is visible in aptitude, and in the intellective potency everything which is intelligible. Therefore, because the inclination to act is in the aptitude, both the visual and the intellectual potency are inclined to act toward the universal, as toward something naturally understood as good. It follows, then, that the soul was not addressing itself to the deaf or the blind, when it counseled its thoughts to curb the sight; for although the sight may not be the proximate cause of desire, it is nevertheless the primary and underlying cause of it.

CIC.   What do you mean by this last statement?

TANS.   I mean that it is not the sensible or intelligible appearance of a form or species which of itself moves the soul, for he who contemplates the form as it is manifest to the eyes does not yet come to love it; but from the instant when the soul conceives the form as an object no longer of sight but of thought, no longer divisible but indivisible, no longer under the species of a particular thing, but under the species of the good and the beautiful, then at once love is born. Now this is the object from which the soul would divert the eyes of its thoughts. This sight is wont to encourage the inclination to love more than it sees; for as I said a little while ago, the affection always considers—by its universal knowledge of the beautiful and the good—that beyond the species of the good and the beautiful, which it has been able to attain, there are infinitely more and more species.

CIC.   But how does it happen that having abstracted a species of beauty which is a conception of the soul we still desire to feed upon its external appearance?

TANS.   Because the soul always desires to love more than it loves and to see more than it sees. Moreover the soul desires that this species which the sight has engendered in it should not become attenuated, enfeebled or lost. The soul therefore wishes to see even more and more, so that what might become darkened to the soul's internal affection might be frequently illumined by the external aspect of the species, which, having been the beginning of its existence ought to be the beginning of its conservation. A similar analogy exists between the act of seeing and the act of understanding, for the sight is proportioned to visible objects exactly as the intellect is proportioned to intelligible objects. I believe, then, that you now understand the intention and sense of the words the soul speaks when it says, *curb your sight.*

CIC.   I see very well. Now proceed to relate what comes of these thoughts.

TANS.   There follows the complaint of the mother against her sons who, having opened their eyes and fixed them upon the splendor of the object, contrary to her command, now wander in the company of the heart. Thus she says:

And you, cruel sons, you abandon me to embitter my pain the more; and because you constantly oppose me, you carry off with you my every hope.

For what reason do I remain conscious, oh covetous heavens? For what reason are these powers mutilated and wasted, if not to make of me the subject and example of so heavy a martyrdom and of so long a punishment?

Oh, in the name of God, dear sons, let even my winged fire become a prey, and let me see some one of you again

returned to me from those tenacious claws. Alas, no one returns, a tardy consolation for my woe.

Here am I miserable, deprived of a heart, abandoned by my thoughts, bereft of the hope I had entirely placed in them. Nothing else remains but the sense of my poverty, unhappiness and wretchedness. And why am I not deprived of this sense too? Why does death not come to my aid, now that I am deprived of life? For what purpose are my natural faculties deprived of their power? How shall I be able to feed upon the intelligible species alone, the food of the   intellect, if my substance is a composite? [15] How shall I be able to remain in the company of these dear and friendly members, which I have woven around myself; how shall I order them according to the symmetry of their elements, if I am abandoned by my thoughts and passions because they are intent upon immaterial and divine food? Come, come, oh

---

[15] i., e., of body and soul.

my fleeting thoughts, my rebellious heart. Let the
sense live on sensible things and the intellect upon
intelligible things. Let matter and the corporeal subject be the support of the body, and the intellect be
satisfied by its own objects; so that this complex continue to subsist, so that there be no dissolution of this
machine, whose spirit unites the soul to the body.
Why, wretched that I am, rather by my own doing
than through external violence, do I witness this hor-
rible divorce within my parts and members? Why?
Because the intellect meddles by ruling the sense and
depriving it of its nourishment; the sense, on the
contrary, resists the intellect, for it would live according to its own rules, and not according to those
of the other. Only its own rules and not those of the
other can assure its existence and its happiness, because it must care for its own and not the other's convenience and life. There are no harmony and concord
where there is that uniformity whereby one nature
wishes to absorb the whole being; but harmony and
concord are present where there is order and due
proportion among diverse things and where each thing
serves its own nature. Therefore let the sense feed
itself according to the law of sensible things, the flesh
according to the law of the flesh, the spirit according
to the law of the spirit, the reason according to the
law of the reason; let them not be confused or troubled with one another. It suffices that one does not at
all alter or prejudice the law of the other. For if it is
unjust that the sense outrage the law of reason, it is
equally blamable that the reason tyrannize over the
law of the senses, inasmuch as the intellect is the
greater wanderer and the sense more domestic and as
though in its own abode.

This is why it is then, oh my thoughts, that some
of you are obligated to care for your home, while
others can set out to seek other cares elsewhere. Such
is the law of nature and such consequently is His law
who is the author and the principle of nature. Therefore
you transgress when, seduced by the beauties of the
intellect, you leave the other part of me in danger of
death. Whence have you engendered this perverse and
melancholy humour of breaking certain natural laws
of the true life, a life you hold in your power, for an
uncertain life that is nothing if not a shadow beyond

the limits of the imaginable? Does it seem natural to you that creatures should refuse the animal or the human life in order to live the divine life when they are not gods but only men or animals?

It is a law of fate and of nature that each thing work according to the condition of its nature. Why, therefore, in pursuit of coveting the nectar of the gods do you lose that nectar which is proper to you, afflicting yourself perhaps with the vain hope of some other nectar? Do you not believe that nature should disdain to accord you this other good, when you so stupidly disdain the good she offers you?

> Heaven scorns giving a second good
> To one who has not held first one dear. [16]

By these and similar arguments the soul, pleading the cause of its more infirm part, seeks to recall the thoughts to the care of the body. But those, although late, return and show themselves to it not in the form in which they formerly departed; they return only to declare their rebellion and to force the whole soul to follow them. That is why the soul utters the dolorous complaint:

> Oh, dogs of Actaeon, oh ungrateful beasts,
> whom I had directed to the refuge of my goddess,
> you return to me void of hope; and coming to the
> maternal shore,

> too grievous a pain do you bring back. You
> tear me to pieces and wish me deprived of life.
> Then leave me, life, become a double stream de-
> prived of its source, that I may reascend to my sun.

> When will nature agree to release me of my
> grievous burden? When will it come to pass that
> from here I too may raise myself ·

> and swiftly be delivered to the lofty object
> and together with my heart and common offspring
> dwell there? [17]

---

[16] LUIGI TANSILLO, *Il vendemmiatore* (Napoli, 1893), p. 60.
[17] See above, introduction, p. 37.

The Platonists hold that with respect to its superior part the soul consists only in the intellect, so that it is more reasonably called intelligence than soul; for it is called soul only in so far as it vivifies the body and sustains it. [18] Therefore here the same essence which nourishes the thoughts and maintains them on high in the vicinity of the exalted heart experiences a sadness in its inferior part and recalls those thoughts as rebels.

CIC. So that there are not two contrary essences, but only one essence subject to two extremes of contrariety?

TANS. Exactly. As the ray of the sun reaches the earth and touches the inferior and obscure elements it illuminates, vivifies and enkindles, but is for all this no less in contact with the element of fire, that is, with the star whence it proceeds, is diffused and has its principle and own original subsistence, similarly the soul which is in the horizon of its corporeal and incorporeal nature, raises itself to superior things and inclines to inferior things. [19] And you can see that this happens not by reason and order of local motion, but only through the impulse of the one and the other potency or faculty. For example when the sense mounts to the imagination, the imagination to the reason, the reason to the intellect, the intellect to the mind, then the whole soul converts itself to God and inhabits the intelligible world. From there by a contrary conversion the soul descends to the sensible world by the degrees of the intellect, the reason, imagination, sense and the vegetative faculty. [20]

CIC. Indeed, I have been told that the soul that finds itself in the ultimate degree of divine things, justly descends to the mortal body and from there climbs again the divine degrees; [21] and also that there are

---

[18] PLOTINUS *Enneads* 4, 8, 4.
[19] MARSILIO FICINO, *Opera Omnia* (Basel, 1561), pp. 657-658:

> ...Anima rationalis quemadmodum omnes ibi convenimum in orizonte, id est, in confinio aeternitatis et temporis posita est, quonia inter aeterna et temporalia naturam mediam possedet, et taquam media rationalis vires actionesque habet ad aeterna surgentes, habet quoque vires operationesque declinantes ad temporalia...

[20] FICINO, *Opera*, p. 119.
[21] PLATO *Phaedrus* 247-250.

three degrees of intelligences—those in which the intellectual dominates over the animal, called celestial intelligences; those in which the animal prevails over the intellectual, called human intelligences; and others in which the two balance each other as in the intelligences of demons or heroes.

TANS.

In exercising its faculty, then, the mind can desire an object only to the extent that it is near, proximate, known and familiar to it. Thus a pig cannot wish to be a man nor desire anything appropriate to the appetite of a man. He prefers to wallow in the mud rather than in a bed of fine linen; he would sooner mate with a sow than with the most beautiful woman nature produces, because the desire conforms to the nature of the species. And among men one can see it is the same, according as some men are more or less similar to one or another species of brute animals. Some men have something of the quadruped, others something of the volatile animals and perhaps these men have an affinity—one I would not wish to describe—which draws them to the love of certain kinds of beasts. Now, if the mind, finding itself oppressed by the soul's tie to the body is permitted to raise itself to the contemplation of another state which the soul can attain, it certainly will be able to see the difference between one state and the other, and to disdain the present for the sake of the future one. Similarly, if a beast were sensible of the difference between his own condition and that of man, between the state of his own ignobility and the nobility of the human state which he would not deem impossible to achieve, then, as a way out, he would prefer death to a life that would detain him in his present existence. Therefore at this point when the soul laments, saying, *O dogs of Actaeon*, it is introduced as something constituted only of the inferior potencies, and the mind has revolted against it, and carried the heart away, that is, it has carried away all the affections and the entire army of thoughts. For that reason, perceiving its present state, and in ignorance of any other, believing none other any longer exists, and having no knowledge of it, the soul laments that its thoughts, in their tardy return, come back rather to draw it up with them than to find any refuge in it. And because of the distraction it suffers from the double love of material and intelligible things, the soul feels itself lacerated and torn to pieces, so that it must finally yield to the more vigorous and powerful attraction. Now if the soul ascends by virtue of contemplation, or is transported above the horizon of the natural affections, perceiving with a most pure eye the difference between the life of contemplation and the life of passion, then, conquered by its most lofty thoughts, as though dead to the body, it aspires to the superior regions; and although it continues to live in the body, the soul vegetates there as if dead and is present in the body as an animate potency incapable of any action; not that it is inoperative so long as the body exists, but that the operations of the soul as a composite are delayed, enfeebled, and debilitated.

CIC.   This, then, is the sense in which a certain theologian, who is said to have been transported to the third heaven, was dazzled by the heavenly vision, and desired the dissolution of his body. [22]

TANS.   In this manner, although the soul at first launches complaints against its heart and thoughts, it now desires to be raised with them and manifestly deplores the union and familiarity contracted with corporeal matter. *Leave me then*, it cries, corporeal life, and do not trouble me, so that *I may reascend* to my native home, *to my sun*. From now on leave me to dry the tears from my eyes, eyes I can no longer aid, separated as I am from my good. Leave me, for it is neither proper nor possible for a double stream to flow *deprived of its source*, that is deprived of its heart; for how can I form two rivers of tears here below, if my heart, the source of those rivers, has flown above with its nymphs which are my thoughts? Therefore, little by little from its disaffection and regret the soul progresses toward a hatred of inferior things which it expresses by the words, *When will nature agree to release me of my grievous burden?*

CIC.   I understand this very well, and even what you would infer with respect to the principle point of this discourse, that there are degrees of loves, affections and frenzies, according to the degrees of greater or lesser light of cognition and intelligence.

---

[22] II *Cor* 12: 2:

...I know a man in Christ above fourteen years ago (whether in the body I know not, or out of the body, I know not; God knoweth), such a one caught up to the third heaven...

TANS.   You understand me well. This should lead you to that doctrine commonly borrowed from the Pythagoreans and the Platonists according to which the soul makes the double progress of ascent and descent, corresponding to the double concern it has for itself and for matter, inasmuch as it is moved by the appetite for its proper good on the one hand, and as its material part on the other hand is directed by the providence of fate.

CIC.   But please tell me briefly what you think about the world soul. Can it too ascend and descend?

TANS.   If you speak of the world as the vulgar refer to it, when they call it the universe, I reply that this world being infinite and without dimension or measure appears to be immobile, inanimate and unformed, even though it is the place of an infinite number of movable worlds and has infinite space in which are all those large animals we call stars. If you speak of the world according to the meaning held among the true philosophers for whom the world is every globe, every star, this our earth, the sun's body, the moon and even others, I reply that the soul of each of these worlds not only ascends and descends but moves in a circle. Because each of these souls is composed of superior and inferior powers, the superior powers lead it toward the divinity, the inferior ones toward the material mass which becomes vivified by that divinity and maintained among the tropics of generation and corruption of the living things of these worlds; and each soul eternally serves its own life; and the action of divine providence always in the same measure and order, by warmth and divine light always maintains it in the same, customary state. [23]

CIC.   This suffices me on this subject.

TANS.   Just as these particular souls according to the diverse degrees of their ascent and descent are diversely affected in their behavior and inclinations, so they manifest a diversity of manner and degree of frenzy, love and sensitivity; and there is this diversity not only in the ladder of nature according to the order of the diverse lives the soul assumes in diverse bodies as expressly held by the Pythagoreans, the Saducees

---

[23] For the relation between the world soul and its body see PLOTINUS *Enneads* 4. 3, 9-11.

and others and implicitly by Plato [24] and those who
have more profoundly penetrated his meaning, but
also in the ladder of human affections which has as
many degrees as the ladder of nature, inasmuch as
man in all his potencies represents every species of
being.

CIC. For that reason souls can be known to ascend or
descend by their affections, to come from above or
from below, to be on the way of becoming beasts or
gods, according to their specific natures, as the Pythagoreans understood it. Or one may understand it
simply by the similitude of the affections held by common opinion; for the human soul need not have the
power to become the soul of a brute, as Plotinus and
other Platonists justly maintain, following the lesson
of their master.

TANS. Good. Now, to come to the point, this soul of
which we speak having advanced from an animal to
an heroic frenzy, expresses itself in these words: *When
will it come to pass that I raise myself to the lofty
object, and dwell there in the company of my heart
and common offspring?* It continues with the same
proposal when it says:

> Destiny, when shall I be allowed to ascend
> the mount, which for my perfect blessing shall
> bring me to the lofty gates where I shall know
> those rare beauties? When will my tenacious
> pain be strongly comforted
>
> by him who reassembles my dislocated mem-
> bers and preserves my failing powers from death?
> My spirit will prevail over its enemy, if it
> ascends where error assails it no longer,
>
> and attains the end it waits for, and
> ascends where the lofty object is, and seizes
> the good which one alone possesses,
>
> whereby so many faults are remedied and
> happiness is found—as he declares who alone
> predicts all things.
>
> Oh *destiny*, oh fate, oh divine and immutable prov-
> idence, *when shall I be allowed to ascend that mount,*

---

[24] PLATO *Phaedrus* 247-249.

when will I reach so much loftiness of mind that I may transport myself and reach those high portals and enter to see those *rare beauties,* beauties that in some way shall be explained and understood? When will he accord efficacious comfort to my pain (releasing me from the rigorous knots of care), *he who reassembles and unites my members,* till then *disunited and dislocated?* The question is asked of Love, who brings about the union of these corporeal members, till then divided from each other as much as one contrary is divided from another; of Love, who, besides, preserves from death these intellectual potencies which have been failing to act, and provides them with the spirit whereby they may aspire to ascend. When, I say, shall I be fully comforted by giving these potencies free flight, so that my whole substance can fix its home in that place where by my own effort I may amend all my faults? Arriving at that summit *my spirit will prevail over its enemy,* for nothing is present there that may outrage it, no contrary that may conquer it, no error that may assail it. Oh, if my spirit *attains* and reaches the place which with all its power it desires, if it climbs and arrives at the summit where its *object is* and settles itself to remain there; if it manages to possess the good which cannot be possessed except by one alone (that is, by that good itself, inasmuch as everything else has goodness only in the measure of its own capacity, and that good alone has it in all its plenitude), then I shall be permitted to be happy according to the mode in which *he declares who predicts all things,* that is, he who declares this loftiness and in whom declaring and accomplishing are the same thing. I will be happy according to the way in which he declares or acts, who predicts everything; that is to say, he who is the principle and efficient cause of all things, for whom to declare and to order is the true making and undertaking. This is how Love's affection makes its way from above and from below upon the ladder of superior and inferior things, and how the intellect and the sense make their way from above and from below in the order of intelligible and sensible things.

CIC.   Therefore the greater number of philosophers hold that nature delights in the vicissitude which is seen in the revolution of its wheel.

END OF THE FOURTH DIALOGUE

I.

CIC.   Let me have a look here, so that by my own effort I may be able to consider the states of these frenzies, according to the arrangement of the militia presented here.

TANS.   Notice how the warriors carry the emblems of their affections and their fortunes. Let us consider their names and their dress. Let it suffice us to give our attention to the meaning of the emblems and to the meaning of what is written, as well as to the motto which accompanies the emblematic figure and the poem which completes the figure by clarifying its sense.

CIC.   This is most agreeable. Here then is the first one. He carries a shield divided in four colors; on the crest of the shield is painted a flame underneath a head of bronze, from whose apertures a smoky wind issues with great force and written above are the words, *At regna senserunt tria.* [1]

TANS.   I shall give you some clarification of the above. As one can see, the presence of the flame warms the globe, in which water is contained, and causes this humid element, rendered lighter and less dense by virtue of the heat, to resolve itself into vapor and consequently to demand a much greater space to contain it. If the water does not find an easy exit, it bursts forth with the greatest force and destruction to crack the vessel; but if an easy exit is procured for it, it issues out little by little with less violence and according to the extent of its evaporation exhales and

---

[1]   "But three realms afflict him." See above, introduction, p. 53 and note 18.

expands into air. This figure represents the frenzied one's heart whose organization has been well disposed to the contact of love's flame, and consequently from its vital substance one part (of the heart) [2] sparkles in flames, another part is transformed into abundant weeping rising from the breast, and still another sends up a wind of sighs to incense the air.

And that is the reason for the words, At regna senserunt tria. Here the word at has the virtue of implying difference, diversity and opposition, as if to say that there is some one else who is capable of experiencing the same feelings, and yet does not experience them. [3] This is very well explained in the verse placed underneath the emblematic figure:

> From my twin lights I, a little earth, am
> wont to pour forth no sparing humour to the sea;
> the sighs hidden within my breast the avid winds
> receive in no small measure;

> and the flame loosed from my heart mounts
> to the sky without diminishing. With tears,
> sighs and my ardor I render a tribute to the sea,

> to the air and to the fire. [4] Water, air
> and fire receive some part of me; but my goddess
> shows herself so iniquitous and cruel,

---

[2]   Parentheses mine.

[3]   For the statement that "some one else" is a reference to the beloved object, see Argument of the Nolan, p. 69 and note 14.

[4]   PETRARCH Rime 35:

> Solo e pensoso i più deserti campi
> Vo mesurando a passi tardi e lenti,
> ... ...
> Si ch' io mi credo omai che monti e piagge
> E fiumi e selve sappian di che tempre
> Sia la mia vita, ch' è celata altrui.

> Alone and thoughtful the more deserted field
> With footsteps dull and slow, I tread
> ... ...
> So that now I believe that mountains and hills
> And rivers and forests know of what temper
> My life is, which from others is concealed.

that my tears find no solace in her, nor
does she hear my cries, nor does she ever turn
in pity toward my ardor. [5]

Here the material subject represented by the earth
is the substance of the frenzied lover. *From twin
lights*, that is to say, from his eyes, he pours forth
copious tears which flow into the sea; from his breast
he sends an abundance and multitude of sighs to the
immense receptacle of the air; and the fire of his heart
does not abate upon the stream of air like a small or
weak flame, does not resolve into smoke and transmi-
grate into another essence, but, powerful and vigorous
(rather nourishing itself upon some other substance
than abandoning anything of its own), it joins a kindred
sphere.

CIC.   I have understood it well. Now to the other.

II.

TANS.   He who comes next has on his shield, also divided
into four colors, a crest in which the sun extends its
rays upon the back of the earth; and there is the motto, *Idem semper ubique totum*. [6]

CIC.   I see that this cannot be easy to interpret.

---

5 Dante *Vita Nuova* 14:

    Coll' altre donne mia vista gabbate,
    ... ...

    Se lo saveste, non potria pietate
    Tener più contra me l'usata prova;
    ... ...

    With the other ladies you mock at my aspect,
    ... ...

    If you knew it, pity would no more hold
    Her usual obduracy against me;
    ... ...

6. "Always and everywhere the same." Théodore de Bèze employs an
emblem of three concentric spheres of earth, moon and sun. The sun is the
largest circle and represents Christ. The moon's circle represents the man
who has died in Christ and now enjoys the direct reflection of His radiance;
the earth's circle represents the man on earth without faith, who, because he
is so remote from the sun, believes erroneously that the moon is actually
darkened by the sun's overwhelming radiance. See Théodore de Beze, *Les
Vrais Portraits des hommes et quarantequatre emblèmes Chrestiennes* (n. p.,
1581), emblème 52, p. 282.

TANS.   The meaning is the more excellent, as it is the less vulgar, and you will see that it is single, unified and not strained. [7] You must consider that although the sun appears different with respect to different regions of the earth according to time and place, nevertheless with respect to the entire globe it acts always and everywhere in the same way, for in whatever point of the ecliptic it may find itself, it causes winter, summer, autumn and spring, and the entire earthly globe receives these four seasons because of it. For it is never hot in one part but it is cold in an other. When it is hottest for us in the tropic of Cancer, it is coldest in the tropic of Capricorn, so that the sun is the cause of the summer here, the winter there, and the cause of the spring and autumn according to the disposition of the middle and temperate regions. Therefore the earth is always subject to rain, wind, heat, cold; in fact the earth would not be wet in one part, if it were not dry in the other, and the sun would not heat it from one side, if it had not withdrawn its heat from the other.

CIC.   Before you complete your argument, I understand what you and the frenzied lover mean. As the sun always directs its impressions upon the earth and as the earth always receives all of them entirely, so does the lover's object by its active splendor render him passively to tears, symbolized by the waters, to passions, symbolized by the flames, and to sighs, symbolized by these intermediate vapors which depart from the fire and proceed to the waters, or depart from the waters and proceed to the fire.

TANS.   It is very well explained in the following sonnet:

> When the sun sets in Capricorn, there is
> no torrent the rains do not enrich; when it
> returns through the Equinox, then are unleashed
> the messengers of Aeolus,
>
> and it enkindles us by a more prolific day
> whenever it reascends to burning Cancer. But

---

[7] Bruno follows Paolo Giovio who holds that in a good emblem the symbol, the idea and the motto must combine in a harmonious manner. See Giovio, p. 7-12. Whether or not Bruno is consistently faithful to Giovio's principle is, of course, conjectural.

my tears, sighs and ardors do not accord with
these frosts, tempests and hot seasons;

for I am always in tears, no matter how
intense my sighs and fires. And though I know
too much of water and fire,

never does it happen that I sigh the less,
and there is no limit to my burning amid sighs
and grievous weeping.

CIC.　　The meaning of the emblem is explained less by
this poem than by the preceding commentary; for the
poem follows rather as a consequence and companion
of the commentary.

TANS.　　Say rather that the emblem is implied in the commentary, and the motto is fully explained in the poem.
For both the emblem and the motto are most appropriately represented by the symbol of the sun and
the earth.

CIC.　　Let us proceed to the third.

III.

TANS.　　The third lover carries upon a shield a nude boy
lying upon the green meadow. The boy rests his head
upon his arm, and turns his eyes to the sky toward
certain edifices, houses, towers, landscapes and gar-
dens set above the clouds; and a castle is also to be
found whose walls are made of fire, with the motto,
Mutuo fulcimur. [8]

CIC.　　What does this mean?

TANS.　　You are to understand that the nude boy represents the frenzied lover, simple, pure and exposed to
all the accidents of nature and fortune, who with his
powerful imagination builds castles in the air and,
among other things, a tower, whose architect is love,
whose walls are the amorous fires and whose builder

---

[8] "Mutually we are sustained". See ANDREA ALCIATI, *Emblematum
flumen abundans,* ed. Henry Green (London, 1871), pp. 115-117, for his
analogous emblem of a nude boy playing a lyre as he rides a dolphin in
the sea against a background of a castle on a rocky cliff. Alciati's boy represents the lover who struggles with love's passions. For William Shakespeare's allusion to Sebastian as Arion on the dolphin's back see *Twelfth
Night,* I, ii, 15-17.

is himself who says, *Mutuo fulcimur*. This is to say, I build and sustain you up there with my thoughts, and you sustain me here below with hope. You would not exist were it not for my imagination and my thought which forms and sustains you; and I would not be alive were it not for the consolation and the comfort I receive because of you.

CIC.   It is true that even the most vain and chimerical fancy can be a more real and genuine medicine to a frenzied heart than the herbs, stones, oils or other products produced by nature.

TANS.   Magicians can do more by means of faith than doctors by means of the truth, and in the gravest illnesses the sick profit more by believing all that the first say, than by understanding all that the second do. Now let us read the verse.

> Beyond the clouds, in the highest region,
> sometimes when I burn in delirium, for the refreshment and deliverance of my spirit I form
> a castle of fire in the air.

> If my fatal destiny incline a little, so
> that the sovereign grace bend without scorn and
> anger toward the flame which kills me, oh happy
> my pain and my death!

> Oh, youth, of your flames and of your
> snares—because of which men and gods sigh
> and become slaves—

> I do not feel the ardor, nor the burden,
> but, you, oh love, can cause them to possess me,
> if your merciful hand will lead you to uncover
> my torment.

CIC.   The lover in this poem shows that what nourishes his fancy and revives his spirit is the belief (for he lacks the boldness to explain and make known his pain to himself, profoundly subject as he is to martyrdom) that, if severe and rebellious fate bend somewhat (and finally decide to smile upon him) by making the lofty object reveal itself to him without scorn and anger, such good fortune would make him deem no joy so happy, no life so blessed as the happiness he would find in his pain and the blessedness he would find in death.

TANS.   And thus he begins to explain to Love that, if it can ever have access to his heart, it will never be by using the armed might whereby he usually triumphs over men and gods; but only by uncovering his burning heart and tormented spirit; for only by such a sight will compassion be able to open the way to him and introduce him to that difficult abode.

IV.

CIC.   What is the meaning of that fly which flies around the flame and is almost at the point of being burned, and the meaning of the motto, *Hostis non hostis?* [9]

TANS.   It is not difficult to understand that the fly, which is seduced by the beauty of the dazzling light, throws itself innocent and full of love into the deadly flame. For that reason *hostis* refers to the scalding effect of the flame; *non hostis* refers to the desire of the fly. Thus *hostis*, the fly as passive; *non hostis* (the fly) [10] as active. *Hostis,* the flame because of its fire; *non hostis,* because of its splendor.

CIC.   Now what is that written on the tablet?

TANS.   May it never be that I lament of love, without which I do not wish felicity. Even if it be true that I toil for it in pain, I can only desire what it grants me.

Whether the sky is clear or obscured, cold or burn-
[ing,
I shall ever be a true phoenix, for another destiny or fate can hardly untie that knot which death cannot untie.

For the heart, for the spirit and for the soul there is no pleasure, liberty or life which smiles so much, rejoices and is so welcomed,

---

[9]   "An enemy yet not an enemy." Bruno's emblem of a fly flying around a flame is a variant of the moth and the sun, among the oldest symbols found in secular and religious literature throughout the middle east and western Europe. The moth represents usually the man who, even at the cost of his own destruction, longs to be one with the infinite. The sun represents a variant of the god-image which may destroy or renew him. Accordingly, Osiris, Tammuz, Adonis, Mithras and Christ were variably adored as sun-gods. See CARL JUNG, *Symbols of Transformation,* trans. R. Hull (London, 1956), pp. 79-109.

[10]   Parenthesis mine.

is so sweet, so gracious and so excellent as the
hardship, yoke and death provided for me by nature,
will and destiny.

This emblem shows the similarity between the
frenzied lover and the fly drawn toward the light.
But then the poem makes apparent their difference
more than their similarity. For one ordinarily believes
that if the fly could foresee its own ruin, it would
rather flee the flame than pursue it as it does now, for
it would hold it evil to lose itself by dissolving in the
inimical fire. But the frenzied one would like to perish
in the flames of love no less than he would like rap-
turously to contemplate the beauty of that rare splendor beneath whose sway by the inclination of nature,
his own free choice and the disposition of fate he toils,
serves and dies, more joyful, more resolved and more
valiant than the influence of any other pleasure offered
to his heart, liberty offered to his spirit and life
reawakened in his soul. [11]

---

[11] PETRARCH *Rime* 19:

Son animali al mondo de sí altera
    Vista che 'n contr' al sol pur si difende;
... ...

Et altri, col desio folle che spera
    Gioir forse nel foco, perché splende,
    Provan l'altra vertú, quella che 'ncende...
... ...

Però con gli occhi lacrimosi e 'nfermi
    Mio destino a venderla mi conduce;
    E so bene ch'i'vo dietro a quel che m' arde.

There are animals in the world of such proud mein,
    That they can defend themselves even against
                                        the sun;
... ...

And others who hope with foolish desire
    Perhaps to delight in the flame because of its
                                        splendor,
    Seek another virtue, that which burns...
... ...

To that end with tearful and infirm eyes
    My destiny leads me to see her;
    And well do I know I follow that which consumes
                        me.

CIC.   Tell me, why does he say, I shall ever be one? [12]

TANS.   Because he thinks it worthy to explain that the reason for his constancy is that the wise man does not change like the moon. It is the stupid man who changes as the moon does, [13] but this lover is one and immovable, like the phoenix. [14]

V.

CIC.   Good. But what does that branch of palm mean, accompanied by the motto, *Caesar adest?* [15]

TANS.   Without too much discussion all may be understood by reference to the writing on the tablet:

Unconquered hero of Pharsalia, [16] although
your warriors were almost extinct when they
saw you, they rose again most potent in battle
and subdued your haughty enemies.

Thus does my good, which is equal to heaven's
blessedness, in revealing itself to the sight of my
thoughts whose light was obscured by my scornful
[soul,
revive them so that they are more powerful than love.

Its sole presence, or the memory of it,
so revives them, that with sway and divine
power

they reduce every contrary violence. My
good governs me in peace, but does not abandon
its snare nor its torch.

The inferior powers of the soul, like a valiant and
inimical army which one finds disciplined, skilled and

---

12   i., e., a true phoenix.
13   Eccl. 27: 12.
14   See below, pp. 154-156.
15   "Caesar is here." Paolo Giovio employs a similar emblem of a branch of palm to the end of which is tied a small marble. In accordance with the political or military purpose of most of his emblems, Giovio's palm represents the powerful reign of the Duke of Urbino, which has since declined. But as it is the nature of the palm to be flexible, so is it the nature of a political reign to rise again. See PAOLO GIOVIO, *Ragionamento sopra imprese* (Venezia, 1566), pp. 51-52.
16   Lucian Pharsalia vii, 539-549.

well provided in its own country, sometimes turn against the foreign enemy, who descends from the high summit of the intelligence to dominate the people of the valley and the swampy plains. [17] It happens that, because of the harassing presence of the enemy and the difficulty of the precipitous swamps, these people find themselves almost lost, and in fact would be lost, were it not for a certain conversion by the act of contemplation to the splendor of the intelligible species; for by the act of contemplation there is a conversion from the inferior to the superior degrees.

CIC.   What are these degrees?

TANS.   The degrees of contemplation are like the degrees of light. Light, which is never in darkness but sometimes appears shadowy, is seen better in colors in the order of their progression from one extreme, black, to the opposite extreme, white; is more efficaciously in the refulgence diffused upon refined and transparent bodies as in the reflection of a mirror or the moon; is more vividly in the rays scattered from the sun, and in the highest and most principal degree, is seen in the sun itself. [18] Now the potencies of comprehension and affection are ordered in such a way that a potency always has an affinity for the one immediately above it, and each potency by a conversion toward the one which raises it reinforces itself against the inferior one that draws it down (as the reason, converted to the intellect, is not seduced or conquered by the sensitive powers); consequently, when the rational appetite clashes with the sensual concupiscence and by the act of contemplation confronts the intellectual light, then it retrieves its lost virtue, reinforces its nerves, frightens the enemy and puts him to rout.

CIC.   In what way do you mean that this conversion takes place?

TANS.   By three preparations which the contemplative Plotinus notes in his book *Of the Divine Intelligence*. [19] The first is by resolving to conform the vision to the divine likeness by turning the sight from things equal or inferior to its own perfection; the second is by

---

[17]   This passage contains the hyperbole and confused metaphor which occur frequently in Bruno's style.

[18]   DANTE, *Paradiso*, 2, 25-97; see also PLOTINUS *Enneads* 5. 8.

[19]   PLOTINUS *Enneads* 5. 8.

applying the vision with every purpose and attention to the superior species; the third is by submitting the entire will and affection to God. For he who behaves in this way is beyond a doubt infused with the divinity, present everywhere and ready to penetrate him who turns himself to it by an intellectual act and offers himself to it by the will's affection without reserve.

CIC. Then it is not corporeal beauty which this lover longs for?

TANS. Certainly not; because not being true or constant, corporeal beauty cannot be the cause of true or constant love. The beauty one sees in a body is an accident and a shadow, and is like other things that are altered, tainted and wasted by the mutation of the subject, which from beautiful often becomes ugly without any alteration taking place in the soul. The reason then apprehends the truest beauty by converting itself to the thing which gives the body its beauty and its form; and this thing is the soul, the modeller and sculptor of the body. After this, the intellect rises further and well understands that the beauty of the soul is incomparably superior to the beauty found in bodies; but it is not persuaded that the soul is beautiful essentially and in itself; for if it were, there would not be the differences one sees within the genus of souls, some of which are wise, amiable and lovely, others stupid, odious and ugly. It is necessary, then, to be raised to that superior intellect which is beautiful in itself and good in itself. This is that one and supreme captain, who alone, placed in the sight of militant thoughts, illuminates them, encourages them, reinforces them and assures them of victory through scorn for every other beauty and the repudiation of every other good. This, then, is the presence which overcomes every difficulty and conquers every violence.

CIC. I understand completely. But what is the significance of, *there it governs me in peace, but does not abandon its snare, nor its torch?*

TANS. It means and proves that love of whatever sort, the stronger its empire and the more certain its power, makes its bonds more tight, its yoke more firm, and flames more ardent, unlike the ordinary prince or tyrant who uses the greatest force and constraint when his power is weakest.

CIC. Let us go to the next one.

VI.

TANS.   Here I see an image of a flying phoenix toward
which a little boy is turned who burns in the midst
of flames, and I see the motto, *Fata obstant*. [20] But
in order to understand this better, let us read the
tablet:

Unique bird of the sun, lovely Phoenix,
who are as old as the world in happy Arabia,
you are still what you always were, while I
am no longer the same.

Because of the fire of love I die unhappy,
while you the sun revives with its rays. You
burn in one, but I in every place. I from
Cupid, but you from Phoebus have your flame.

You have predestined for you the term of
a long life, and I have a brief one, whose end
is offered me in ruins without number.

I know neither the life I shall live, nor
the life I have lived. A blind destiny leads
me, while you, assured of yours, turn once
again toward your light.

The sense of the verse shows us that the emblem
represents the antithesis between the fate of the
phoenix and the fate of the frenzied one, and that
the motto, *Fata obstant*, does not mean the fates are
contrary either to the boy or to the phoenix, or to
the two of them, but that for each one of them the
decrees of fate, far from being the same are different
and opposite. For the phoenix is what it was, inasmuch as by the fire the body of the phoenix is renewed in the same material, and its form is renewed by
the same spirit and soul. The frenzied one is what he
was not, because as a human subject he belonged
previously to some other species, separated from the

---

[20] "Their fates run contrary." Nothing is more common in emblem literature than the phoenix. De Bèze's emblem depicts a phoenix burning
fiercely upón a hill and sending up smoke into the sky. The phoenix represents the Protestant saint who will rise upón his ashes in defiance of the
heretical hangmen who martyred him. See DE BEZE, *emblème* 6, p. 246.

human species by differences without number. Therefore one knows what the phoenix was and knows
what it shall be, but only in terms of many and uncertain metamorphoses shall this lover be able to clothe
himself again in a natural form identical or similar
to the one which is his today. Besides, the phoenix
in the presence of the sun changes death for life, and
this subject in the presence of love changes life for
death. And further, the phoenix consumes itself on the
aromatic altar, and the lover finds his fire everywhere
and takes it with him wherever he goes. Moreover
the phoenix is assured of the terms of a long life, but
the lover because of infinite vicissitudes of time and
innumerable reasons of circumstance has only the uncertain term of a short life. The phoenix enkindles
itself with certainty, the lover burns in the doubt of
ever seeing the sun again.

CIC.   What do you suppose this emblem represents?

TANS.   It represents the difference between the inferior
intellect (commonly called the intellect in potency,
or the possible or passive intellect), which is uncertain, diverse and multiform, and the superior intellect, the one perhaps called by the Peripatetics the
lowest in the hierarchy of the intelligences, which,
they say, immediately influences every individual of
the human species and is the active and actual intellect. [21] This intellect, unique for the human species,
influences every individual and is comparable to the
moon which is always of the same species and whose
aspect ever renews itself as it turns toward the sun,
the first and universal intelligence. [22] However, the
human intellect, individual and multiple, is turned
like the eyes toward countless and most diverse objects, so that it is informed according to an infinity of
degrees and an infinity of natural forms. That is why
it happens that this particular intellect is frenzied,
wandering and uncertain, while the universal intellect
is tranquil, stable and certain with respect to the appetite as well as to the apprehension. Therefore (as
you can easily decipher for yourself), this figure symbolizes the nature of the sensitive appetite and apprehension, changing, shifting, inconstant and uncer-

---

21   ARISTOTLE De anima iii. 4, 5.
22   AVERROËS De anima iii, ed. Juntes (Venice, 1550), p. 165.

tain, and the nature of the intellectual appetite and its concept, firm, stable and definite. The figure also symbolizes the difference between sensual love, uncertain and undiscerning of its objects, and intellectual love which sees only a single object toward which it turns, whereby its thought is illumined, its passion enkindled, inflamed, illuminated and maintained in unity, identity and position.

VII.

CIC.   But what is the meaning of that figure of the sun with a circle inside it and another circle outside of it, and of the motto, *Circuit?* [23]

TANS.   I'm sure I would never have understood the meaning of the figure if the author himself had not explained it to me. Now it must be understood that *Circuit* refers to the motion the sun makes around the double circle drawn inside it and around it to signify that the sun both moves itself and is moved at the same time. Therefore, the sun is always found to be in every point of the traversed circle, for in the single instant of time, it both moves and is moved simultaneously and is equally present in the entire circumference of the circle in which motion and rest converge and become one.

CIC.   This I have understood in the dialogues *Of the Infinite Universe and Innumerable Worlds,* [24] where it is explained that the divine wisdom (as Solomon said) [25] is movable to the highest degree and at the same time most stable, as it is declared and under-

---

[23]   "It revolves in a circle." See above, p. 145 and note 6.

[24]   Bruno combines curiously the title of the Italian work of the London period with the title of his work in Latin. The Latin work, *De immenso et innumerabilibus* was probably in rough draft but not published in 1585. See also BRUNO, *Opere* II, 406 and note 1 and Bruno, *Des fureurs héroïques,* Michel, p. 296 and note 10.

[25]   Wisd. 7: 24.

> ...For wisdom is more active than all active things and reacheth everywhere by reason of her purity...

See also 27:

> ...And being but one, she can do all things, and remaining in herself the same, she reneweth all things...

stood by all those who know. Now proceed to your explanation of it.

TANS. The author of the emblem means that his sun is not like that sun which (as is commonly believed) circles the earth in the daily motion of twenty-four hours and completes its planetary motion in twelve months, affecting the earth by the four distinct seasons of the year according to the regions in which it finds itself in the four cardinal points of the Zodiac. But his sun is such that, representing eternity itself and therefore in perfect possession of all, it comprises the winter, spring, summer, the autumn, the day and the night together, for it is wholly everywhere and in all points and places.

CIC. Now apply your statement to the emblem.

TANS. Because it is impossible to design the whole sun at each point in the circle, two circles have been drawn here. One circle is drawn around the sun to show that the sun moves itself through it. The other circle is drawn inside the sun, to show that the sun is moved by it.

CIC. But this figuration seems to me obscure and not precise.

TANS. It is sufficient that it is as clear and precise as he was able to make it. If you can find a better one, you are given every authority to remove this one and replace it with one of your own. For this was presented only in order that the idea might not be without some concrete form.

CIC. What do you say about the word *circuit?*

TANS. That motto, according to its fullest meaning, represents as much as can be represented; for by the sun's revolving itself and being revolved in a circle is signified its present and perfect motion.

CIC. Most excellent. Granted that those circles express poorly the co-existence of movement and rest, we can nevertheless say that they have been put there to signify a single revolution. And so I am content with the subject and form of the heroic emblem. Now let us read the rime.

TANS.   Sun, you send down temperate rays from
Taurus, from Leo you ripen and burn all, and
when you shed light from stinging Scorpio much
of your fiery vigor you abandon,

until from proud Aquarius you consume
everything with cold, and harden the humid
bodies.—But I in spring, summer, autumn and
in winter am eternally warmed, burned, inflamed
and enkindled.

So hot is my desire, that I am easily
moved to contemplate that lofty object for
which I burn so much,

that my ardor throws off sparks to the
stars. The years have no moment which see
any change in my anguish.

Notice here that the four seasons of the year are
indicated not by the four movable signs of Aries, Cancer, Libra and Capricorn, but by the four which are
called fixed, that is to say, Taurus, Leo, Scorpio and
Aquarius, in order to represent the perfection, stability
and fervor of those four seasons. Note also that by
virtue of those apostrophes found in the eighth verse,
you may read *mi scaldo, accendo, ardo, avvampo;* or,
scaldi, accendi, ardi, avvampi; or also, *scalda, accende,
arde, avvampa.* [26] Besides, you must consider, these are
not four synonyms but four diverse terms which express
so many degrees of the effects produced by the fire;
for first, the fire warms, second, it inflames, third, it
burns, fourth, it enkindles or sets him on fire who has
been warmed, inflamed and burned. And therefore are
denoted in the frenzied one desire, intention, zeal and
the affection of love which he feels at every moment.

CIC.   Why do you give it the name of *anguish?*

TANS.   Because the divine light is in this life more an object of laborious emptiness than of tranquil fruition,
since our minds move toward that light like birds of
the night toward the sun.

CIC.   Let us proceed. I have now heard enough to grasp
everything.

---

[26] The final endings characterize the first person singular of the present
indicative, the second person of the imperative and the third person of the
indicative.

VIII.

TANS.   The following crest presents a full moon with the
motto, *Talis mihi semper et astro.* [27] It means that to
the star, that is, to the sun and to him the moon is always such as it is here, full and clear in the entire
circumference of its circle. So that you may understand
this better, I would have you read the poem written
upon the tablet:

> Inconstant moon, fickle moon, you who
> emerge from the horizon with your horns now
> empty, now full, your orb reascends now white,
> now dark; now you illumine Boreas and the
> valleys of the Caucasus,

> now you turn along your usual path to
> give light to the south and the last confines
> of Lybia. So the moon of my sky for my con-
> tinual torment is ever steady, and is ever full.

> And my sun is the same, which forever
> ravishes and restores me, which ever burns and
> is so resplendent,

> always so cruel and so beautiful. This
> my noble torch ever martyrs me, and still it
> delights me.

It seems to me that this lover's particular intelligence is always thus with regard to the universal
intelligence. In other words, the universal intelligence
illumines the entire hemisphere, even though that intelligence appears sometimes obscure, sometimes more
or less luminous, according to the impressions it makes
upon the inferior potencies. Or perhaps it would
mean that his speculative intellect (invariably in act)
is always turned and drawn toward that human intelligence represented by the moon. For as the moon is
called the lowest among all the planets and is found
nearest to us, so the intelligence which illumines all

---

[27] "Such is it always to me and to the sun." Giovio also employs the
emblem of a full moon. However his emblem characteristically represents
the splendor of Henry II's reign, which gives light and glory to the world.
See Giovio, pp. 20-21.

of us (in our present state) is the lowest in the hierarchy of intelligences, as Averroës and other more subtle Peripatetics note. [28] With respect to the intellect in potency, the human intelligence represented by the moon sometimes seems to decline, insofar as it does not display itself in act, and sometimes it seems to rise from the valley, that is, from the bottom of the concealed hemisphere; sometimes it displays itself vacant and sometimes full, accordingly as it gives more or less light; sometimes its orb is obscure, sometimes brilliant, because sometimes it dispenses only a shadow, similitude and vestige, or sometimes it pours out the light more openly; sometimes it declines toward the south, sometimes to the north; that is, sometimes it retires and alienates itself more and more from us, sometimes it returns and approaches us. But the active intellect by incessant labor (for it is foreign to human nature and the human condition which is wearied, beaten, incited, sollicited, distracted and as though torn by the inferior potencies) always sees its object immobile, fixed and constant and always in plenitude and in the same splendor of beauty. [29] Therefore the object always *ravishes* him insofar as he fails to offer himself to it, and always restores him insofar as he succeeds in offering himself to it. It always enflames his passion as much as it *is resplendent* in his thought; it is always as *cruel* to him by withdrawing itself as he similarly withdraws himself, and always *so beautiful* in communicating itself to the degree that he offers himself to it. *It always martyrs him* separated from him by space; and it always *delights* him because he is conjoined to it in his affection.

CIC.   Now apply the meaning to the motto.

TANS.   He says then, *Talis mihi semper;* that is to say, by means of the constant application of my intellect, memory and will (for they alone do I remember, understand

---

[28] AVERROËS *De anima* iii, ed. Juntes, p. 165. Averroës makes the human intellect the last of the planetary intelligences, an immaterial and eternal form. It is the light that illumines individual souls and makes possible for humanity a recurring participation in the eternal truth.

[29] Averroës holds a passive and active intellect common to all men. The passive intellect does not arise in man but is eternal and individual men participate in it while they think. The mind after death continues to exist, not as an individual substance but as an element of the eternal and universal intelligence. See AVERROËS *De anima* iii.

and desire), it is always such to me, and insofar as I can understand, it is entirely present and is never separated from me by distraction of my thought, never obscured by any deficiency of attention, for there is no thought that turns me from its light, no natural necessity that compels me to attend it less. *Talis mihi semper*, means further that, on its own part, the moon is itself invariable in substance, virtue, beauty, and efficacy with respect to all that shows an invariable constancy toward it. He says, then, *et astro* because with respect to the face of the sun which illumines it, the moon is always equally luminous inasmuch as it is equally turned to the sun and the sun equally diffuses its rays upon it. Although that moon which we see with our eyes appears to this earth sometimes dark and sometimes light, sometimes less brilliant and sometimes more brilliant, it nevertheless receives an equal measure of the sun's illumination, because it always receives the sun's rays at least upon the entire surface of its hemisphere. Similarly this earth is equally illuminated upon the surface of its hemisphere, even though from time to time from its watery area it sends up its light to the moon according to the variability of the light it receives from it. (We think of the moon, as well as each of the innumerable stars, as another earth). Thus both the earth and the moon change their positions toward one another as each one finds itself nearer to the sun.

CIC. How is this intelligence represented by the moon, which shines from its entire hemisphere?

TANS. All the intelligences are represented by the moon, inasmuch as they participate in potentiality and act, and inasmuch, I say, as they have the light unrefined and according to participation because they receive it from another. And these intelligences do not have the light of themselves and by their nature but have it by the view of the sun, the first intelligence, pure and absolute light, pure and absolute act.

CIC. Then everything dependent and not prime act and first cause is as though composed of darkness and light, matter and form, potency and act?

TANS. Exactly. Besides, our soul in its entire substance is symbolized by the moon. It shines through the hemisphere of the superior potencies when turned toward the light of the intelligible world; and it is darkened on the side of the inferior potencies when occupied with the government of matter.

IX.

CIC.   It seems to me the emblem I see on the following shield may contain some issue and symbol relevant to what has already been said. The emblem is a rugged, branchy oak tree blown by the wind and is circumscribed by the motto, *Ut robori robur;* [30] and on the tablet attached to the emblem is the following poem:

> Ancient oak which spreads its branches to the air and fixes its roots in the earth, neither the trembling of the earth, nor the powerful spirits the sky lets loose from the bitter north wind,
>
> nor whatever the dreadful winter may send, can ever uproot you from the place where you stand firm; you demonstrate the true semblance of my faith, which no external accident has ever shaken.
>
> You ever embrace, nourish and contain the same ground in whose depths you spread agreeable roots upon a generous bosom:
>
> Upon one single object I have fixed my spirit, sense and intellect. [31]

TANS.   The motto is clear. The frenzied one is proud that he has the strength and robustness of the oak tree; like one of the lovers before him he is proud to be one and the same with the unique phoenix, and like the one who immediately precedes him, proud to be able to conform to the moon in its everlasting brilliance and beauty. Moreover, he is proud that he does not resemble the moon insomuch as it is variable to our eyes, but insomuch as it always receives an equal measure of the solar splendor. Therefore, he is proud

---

[30] "Strong as an oak." Alciati's emblem depicts a rugged oak tree, almost identical with that of Bruno. The oak tree is compared to Rome's indomitable might, which endures despite the onslaught of her enemies. See ALCIATI, *Emblematum*, p. 49.

[31] For Tansillo's analogous image of a tall and ancient oak see *Poesie liriche*, ed. Fiorentino, p. 76. Fiorentino observes on p. 281 that Tansillo had composed an emblem of an oak tree.

of having remained so constant and firm against the north wind and the tempestuous winters, so strong in the unshakable attachment which fixes him to his sun where his desire and purpose root him, like the oak tree whose roots intertwine with the veins of the earth.

CIC.   For my part I regard it better to remain in peace and free from any onslaught than to find myself in circumstances of such vigorous endurance.

TANS.   There is an aphorism of Epicurus which, if understood properly, would not be judged so profane as the ignorant think it; for it does not deny virtue to be such as I have defined it and takes nothing from the perfection of constancy, but rather adds something to that perfection which the vulgar comprehend; for he believes the true and complete virtue of sturdiness and constancy is not the constancy which resists discomforts and puts up with them, but the constancy which takes them upon oneself without feeling them. He does not hold perfect, divine and heroic the love which feels the spur, the bit, remorse or pain caused by that vulgar kind of love, but heroic that love which abolishes any sense of other affections, so that he attains the degree of pleasure which has no power to annoy him by diverting him or by making him stumble upon some obstacle; and this is to reach the highest beatitude in this state, to have desire and not to have any sense of pain.

CIC.   The common opinion does not accept this interpretation of Epicurus.

TANS.   That is because one does not read his books, nor read those books which report his arguments without prejudice, but those who read the story of his life and the circumstances of his death will understand his meaning in the words he dictated as the exordium to his testament: *Having come to the last and most happy day of our life, we have planned for that day peace, health and tranquillity of mind; for no matter how much, on the one hand, the greatest pain has tormented us with obstacles, that torment, on the other hand, has become completely absorbed by the pleasure we have taken in our creations and in the consideration of our end.* [32] And it is clear that he did not

[32] These words are said to be a liberal translation of a letter of Epicurus to Idomeneus. See BRUNO, *Des fureurs héroïques,* p. 297, and note 18.

find more happiness than pain in eating, drinking, sleeping, and generating. His happiness consisted in feeling no hunger, no thirst, nor fatigue nor sexual appetite. Consider, then, what we hold to be the perfection of constancy. Constancy does not consist in this, that the tree does not allow itself to be shattered, bent or broken; but in this, that it does not even stir. In the likeness of that oak our hero holds fast his spirit, sense and intellect, at that point where no tempestuous onslaught can move him.

CIC.   Do you mean then that to put up with torment is a desirable thing because it is a sign of strength?

TANS.   To put up with *torment,* as you say, is a part of constancy, but it is not its complete virtue; and I call it *putting up with it with hardiness,* and Epicurus calls it, *torment without feeling it.* This privation of feeling results from this that everything has been entirely absorbed in the cultivation of virtue, the true good and happiness. Such was the insensibility of Regulus toward the tomb, of Lucrezia toward the dagger, of Socrates toward poison, of Anaxarcus toward the mortar (which bruised him), of Mucius Scaevola toward the fire, of Horatius Cocles toward the abyss of the Tiber, and of other virtuous men toward the things which greatly torment and horrify those who are ordinary and vile.

CIC.   Now proceed.

X.

TANS.   Look at this other emblem which contains the image of an anvil and hammer and has the motto, *Ab Aetna.* [33] But before we consider it, let us read the poem in which the prosopopoeia of Vulcan is introduced:

To my Sicilian mount where I may temper
the thunderbolts of Jove now I shall not return.
Here I shall remain, I, scabrous Vulcan, for
here a prouder giant rebels,

a giant who is enflamed against the sky and
rages in vain, as he attempts new labors and

---

[33]   From Aetna.

trials. A better forger of Aetna, a better smith,
anvil and hammer do I find

here in this breast which exhales sighs
and whose bellows vivify the furnace, where
the soul lies prostrate from so many assaults

of such long tortures and great martyrdoms,
and brings a concert which divulges so bitter
and cruel a torment.

This poem shows the pains and afflictions inherent
in love, especially in vulgar love, which is nothing else
than the smith's shop of Vulcan who forges the thunderbolts of Jove to torment delinquent souls. For disordered love bears within itself the germ of its own
pain, inasmuch as God is near us, with us and inside
us. There is found in us a certain consecrated mind
and divine intelligence served by a peculiar passion,
the vindicator of the intelligence, which with a certain
remorse of conscience [34] strikes the transgressive soul
as with a heavy hammer. This intelligence observes
our actions and passions, and as we treat it so are we
treated in turn. I say that every lover has his Vulcan,
for there is no man or lover who does not have God
within him. God is most certainly in everyone, but the
kind of god in everyone is not so easily known; and
if it were at all possible to probe the question and
shed light upon it, nothing I believe would clarify it
for us more than love; for love is as one who pushes
the oars, inflates the sail and tempers this composite
(which we are) to the end that it becomes affected
for the better or for the worse.

I say affected for the better or for the worse inasmuch as love operates through moral or contemplative acts, and because there are common afflictions
by which all lovers are wounded. For inasmuch as
things come in mixtures, there is no intelligible or
sensible good to which evil is not joined or opposed,
nor is there any truth to which falsehood is not joined
or opposed; similarly, there is no love without fear,
zeal, jealousy, rancor and the other passions proceed-

---

[34] ξυντηρηδις may be defined as "conscience" in the ethico-religious
sense. See BRUNO, *Opere*, II, p. 13 and note 2.

ing from the one contrary which disturbs us, while the other contrary pleases us. Therefore, as the soul desires to recover its natural beauty, it seeks to purge itself, heal and reform itself; and for this purpose the soul uses fire, for like gold mixed with earth and shapeless, it wishes by a vigorous trial to liberate itself from impurities, and this end is achieved when the intellect, the true smith of Jove, sets to work actively exercising the intellectual powers.

CIC.   This I believe is related to the passage in Plato's *Symposium* where it is said that Love from his mother Penury inherited aridity, leanness, pallor, destitution, submission and homelessness, [35] circumstances which represent the torment of the afflicted soul wearied by contrary passions.

TANS.   It is exactly so; because the spirit affected by this frenzy is distracted by profound thoughts, tortured by pressing cares, burned by fervent longings, and solicited on occasions without number. As a result, because it finds itself suspended, the soul necessarily becomes less diligent and operative with respect to the government of the body and the activity of the vegetative potency. Consequently, the body becomes lean, undernourished, extenuated, deficient in blood and overcome by melancholy humors; and if these humors do not become the instruments of a well disciplined soul and of a clear and lucid spirit, they lead to insanity, to stupidity and to a bestial frenzy, or at least they lead to a negligence of the self and self-disdain which Plato represents by the figure of bare feet. [36] Love becomes debased and flies close to the ground when it is attached to base things; it flies high when it is intent upon the more noble enterprises. In conclusion, then, whatever it may be, love is always afflicted and tortured, so that it cannot avoid becoming material for the furnace of Vulcan; for the soul, a divine thing and by its nature not the slave but the lord of the material body, is thrown into painful disturbance while it voluntarily serves the body where it does not find that which satisfies it. And no matter how much it may fix itself upon the beloved object, the soul cannot avoid being sometimes agitated and

---

[35] PLATO *Symposium* 203, 204.
[36] *Ibid.*, 203.

shaken by hopeful sighs, by fears, doubts, zeal, troubles of conscience, remorse, wilfulness, contrition and other tormentors represented by the bellowings, coals, anvils, hammers, pincers and the other tools found in the work shop of this sordid and squalid spouse of Venus.

CIC.   Now a good deal has been said of this subject. Be so good as to see what follows next.

XI.

TANS.   Here is a golden apple tree most richly enamelled with a variety of the most precious fruits, and this emblem is circumscribed by a motto which says, *Pulchriori detur*. [37] The allusion to the story of the three goddesses who submitted themselves to the judgement of Paris is most familiar. But let us read the verse which will inform us more precisely of the intention of this frenzied one.

> Venus, goddess of the third sphere [38] and
> mother of the blind archer, subduer of all
> men; that other, sprung from the forehead of
> Jove, and the proud wife of Jove, Juno,

> call the Trojan shepherd to judge which
> of them, most beautiful, deserves the golden
> fruit. If my goddess were set among them, it
> would be awarded neither to Venus, Athena or
> Juno.

> The Cyprian goddess is beautiful by reason of lovely limbs, Minerva through her intellect, and Juno pleases by that worthy splendor

> of majesty, which satisfies the Thunderer;
> but my goddess contains within herself all that
> is requisite of beauty, intelligence and majesty.

---

[37] "It shall be given to the more beautiful one." For Alciati's emblem of a fertile tree displaying an abundance of fruit in the spring of the year see Alciati, p. 24.

[38] ANTONIO EPICURO, *Cecaria*, ed. Palmarini (Venice, 1535), fol. C. viii, v⁰. See also PETRARCH *Rime* 302.

In this poem the frenzied one compares his object, which contains and unites the qualities, characteristics and species of beauty to other objects which can only offer one, and, besides, each one distributed among diverse individuals. For example, in the category of corporeal beauty Apollo cannot find every species united in one virgin but distributed among many. Now it happens that here there are three species of beauty, although all three are found in each of the three goddesses; for Venus is not deficient in wisdom and majesty, and Juno is not wanting in beauty and wisdom any more than Athena is wanting in majesty and beauty. Nevertheless, in each of the three goddesses one of these qualities happens to surpass the others and for that reason is considered proper to her, while the other qualities are considered mere accidents; moreover, with respect to the quality which predominates in her, each goddess appears sovereign and outweighs her rivals. And the reason for this difference is that certain qualities do not belong to each goddess primarily and according to its essence, but according to participation and derivation. Just as in all contingent things perfections exist more or less only according to inferior or superior degrees.

But in the simplicity of the divine essence all exists in all and not according to measure; and thus in the divine essence wisdom is not superior to beauty and majesty any more than goodness is superior to power. In fact all the attributes of the divine essence are not only equal, but they are even identical and are one simple thing. In a similar way all the dimensions of a sphere are not only equal (length being equal to depth and breadth) but even identical, because in a sphere that which you call depth you may at the same time call length and breadth. Analogously, in the divine essence the height of wisdom is one with depth of power and breadth of goodness. All these perfections are equal because they are infinite. One must therefore measure the greatness of the one according to the greatness of the other. But where these things are finite, wisdom may surpass beauty and goodness, goodness and beauty may surpass wisdom, wisdom and goodness may surpass power and power may surpass both goodness and wisdom. But where there is infinite wisdom that wisdom can not exist without infinite power, otherwise that wisdom would not possess the power to know infinitely. Where there is infinite goodness that goodness must have infinite wisdom, otherwise that goodness would not know how to be infinitely good. Where there is infinite power that power must also have infinite goodness and wisdom, for the infinite power must have power to know as well as the knowledge of power. You see, then, how the beloved object of this frenzied one who is inebriated with drinking the divine nectar is incomparably higher than any other object. You see, I mean to say, how the intelligible species of the divine essence possesses the perfection of all the other species in the highest degree, so that the degree of participation in the form he can attain will give him the appropriate degree of potential comprehension and action and the appropriate degree of love for this single beauty and disregard and disdain for every other. Therefore, to that one alone who is all in all must the golden apple be consecrated; and it must not be consecrated to beautiful Venus whom Minerva surpasses in wisdom and whom Juno surpasses in majesty; not to Athene whom Venus surpasses in beauty and Juno in majesty; not even to Juno, who is neither the goddess of intelligence nor of love.

CIC.   Certainly just as there are degrees in nature and in essences, so are there degrees of intelligible species and degrees of magnanimity in the affections and frenzies of love.

XII.

CIC.   The following emblem has a head with four faces which blow toward the four corners of the sky. Four winds issue from that single head, and above those winds two stars are seen to rise. The emblem bears the motto, *Novae ortae Aeoliae*.[39] I should like to know what this means.

TANS.   It seems to me that the sense of the emblem follows that of the one just preceding; for, as the former emblem presented an infinite beauty as the object of love, this one presents a very great aspiration, zeal, affection and desire for that infinite beauty; for that

---

[39] "A new Aeolus is born." Alciati's emblem depicts a bearded man with two faces looking in opposite directions to suggest the virtue of prudence. The meaning of the verses in Latin is not clear, but suggests possibly that the prudent man learns from the past in order to provide for the future. See Alciati, p. 24.

reason I believe these winds are meant to represent sighs, as we shall understand if we look and read the verse:

> Zephyrs of the Titan Astraeus and of Aurora, who trouble the sky, the sea and the land, as if discord had hurled you forth into space for having made proud war against the gods,

> you no longer make your home in the Aeolian cave, where my power refrains and bridles you, but are confined within that breast I see constricted by so much sighing.

> Turbulent cohorts of the tempests of one and the other sea, nothing else avails to assuage you

> but those murderous and innocent lights. Those lights when clear, will render you tranquil; when dark, will render you bold.

It is easy to see that Aeolus is introduced as speaking to the winds, which he says are no longer governed by him in his cavern, but are now governed by two stars in the breast of the frenzied one. Here the two stars do not represent the two eyes of a beautiful face but the two intelligible species of the divine beauty and goodness of the infinite splendor which influence the intellectual and rational desire and cause it to aspire infinitely to the extent that it understands the grandeur, beauty and infinite goodness of that excellent light. For if love is finite, content and fixed upon a certain limit, it will not approach the species of divine beauty but a species other than the divine beauty; but if love aspires higher and higher, one may say that it will expand toward the infinite.

CIC.   How can the aspiration be appropriately represented by puffing out? How is desire symbolized by the winds?

TANS.   He among us who aspires to this state, sighs, and also puffs out. And therefore the vehemence of aspiration is conveyed by that hieroglyphic of a powerful puffing out.

CIC.   But there is a difference between sighing and puffing out.

TANS.   The one is not meant to be identical to the other. There is only a similarity between them.

CIC.   Then proceed with your argument.

TANS.   The infinite aspiration, then, expressed by the sighs and symbolized by the winds is not under the government of Aeolus in the Aeolian caves but is under the government of the lights that are mentioned, which murder the frenzied one not only by their innocence but by their supreme benignity, for they make him die to all other things because of his zealous affection. Moreover, if these lights go out or conceal themselves, they render a tempest within him, and, if they are clear, they render him tranquil. Similarly, in a season when a veil of clouds darkens the eyes of the human body, then the zealous soul feels only turbulence and affliction; but if the veil is torn and thrown aside, the soul will enjoy a tranquility noble enough to satisfy its nature.

CIC.   But how can our finite intellect pursue an infinite object?

TANS.   By the intellect's infinite potency.

CIC.   A vain potency if it must remain unfulfilled.

TANS.   The intellectual potency would be vain, if it moved toward a finite act in which its infinite potency would remain in privation; but it would not be vain if it moved toward an infinite act in which its infinite potency enjoys perfect fulfillment.

CIC.   If the human intellect and action are finite by nature, how and why is the intellect endowed with infinite potency?

TANS.   Because it is eternal and because its delight is not limited by time, it knows no end or limit of delight; and because, although finite in itself, it is infinite with respect to its object.

CIC.   What is the difference between the infinity of the object and infinity of the potency?

TANS.   The potency is finitely infinite, and the object is infinitely infinite. But to return to our discourse. The motto says, *Novae ortae Aeoliae* because we may believe that all the winds enclosed in the deep caves of Aeolus are converted into the lover's sighs, if we consider that these sighs are caused by the affection which ceaselessly aspires to the supreme good and infinite beauty.

XIII.

CIC.

Next let us see what the meaning is of that burning torch whose motto is, *Ad vitam, non ad horam.* [40]

TANS.

It signifies the perseverance in love and the burning desire for the true good in which the frenzied one burns while in this temporal state. This, I believe, is what the following tablet teaches:

The peasant leaves his lodging when the
day breaks from the bosom of the Orient, and
when the sun strikes more intensely, tired and
smitten by the heat he sits down in the shade.

Then he works and tires himself until a
dark gloom covers the hemisphere; then he rests.
But I am exposed to continual blows morning,
noon, evening and night.

Those fierce rays which issue from the
two arcs of my sun (as my destiny wills) from
the horizon of my soul

never depart, burning my afflicted heart
at every hour from its meridian. [41]

CIC.

This verse interprets the emblem in a general way without explaining its meaning in detail.

TANS.

And I do not have to strain to show you its precise meanings, for these can be understood if you give them a little consideration. *The sun's rays* are the forms whereby the divine beauty and goodness are manifest to us; and they are *fiery* because they cannot be apprehended by the intellect without consequently enkindling the desire. *The two arcs of the sun* are the two species of knowledge called by the scholastic theologians *matins and vespers;* [42] so that the intelligence which illumines us through the medium of

---

[40]   "For always, not just for an hour."

[41]   For an analogous allusion to the journey of the laborer compared to the lover's plight see PETRARCH *Rime* 50, st. 2.

[42]   For the statement that "morning and evening knowledge" represents knowlege of things as they are primordially in the *Word*, and things as they become after they are created by the Word, see THOMAS AQUINAS *Summa theologica* I. 58. 6.

the air leads the species to us, either in virtue of our admiration of it for itself, or of our admiration for the efficacy contemplated in its effects. *The horizon of the soul* is the region of the superior potencies; and in this region the valiant intellectual apprehension is aided by the vigorous impulse of the affection represented by the heart, which is afflicted because it burns at every hour; for all the fruits of love we can gather in this (mortal) state are not so sweet that they are not mingled with certain affliction; at least the affliction that comes from the consciousness of fruition without plenitude. This is particularly the case in the fruits of natural love, whose condition I should not know how to express better than the Epicurean poet has:

Ex hominis vero facile pulchroque colore
Nil datur in corpus preater simulacra fruendum
Tenuia, qaue vento spes captat saepe misella.
Ut bibere in somnis sitiens cum quaerit, et humor
Non datur, ardorem in membris qui stinguere possit;
Sed laticum simulacra petit frustraque laborat
In medioque sitit torrenti flumine potans:
Sic in armore Venus simulacris ludit amantes,
Nec satiare queunt spectando corpora coram,
Nec manibus quicquam teneris abradere membris
Possunt, errantes incerti corpore toto.
Denique cum membris conlatis flore fruuntur
Aetatis; dum iam praesagit gaudia corpus,
Atque in eo est Venus, et muliebria conserat arva,
Adfigunt avide corpus iunguntque salivas
Oris et inspirant pressantes dentibus ora,
Nequicquam, quoniam nihil inde abradere possunt,
Nec penetrare et abire in corpus corpore toto. [43]

---

[43] LUCRETIUS *De rerum natura* iv. 1094-1111:

...The body is given nothing to enjoy by a pretty face or a pleasant complexion but tenuous images which all too often fond hope scatters to the wind. When a thirsty man tries to drink in his dreams, the liquid which can quench the fire in his limbs is not given him. But he seeks images of spring water with fruitless effort and thirsts nightly in the midst of torrential rivers. Even so in the midst of love Venus mocks her lovers with images, for they cannot satisfy their sight by looking upon her bodily form, nor can they snatch anything of her tender limbs with their hands, as they wander aimlessly over her whole body. Finally they pluck the fruit of life with their joined limbs. But even while their bodies

Similarly does that wise Hebrew judge the manner in which we can enjoy divine things here below.
As we force ourselves to penetrate and unite with
those divine things, we find we are more afflicted by
our desire for them than pleased by our conception
of them. And therefore that wise Hebrew [44] could say
that he who increases wisdom increases pain, for the
greater comprehension nurtures the greater and loftier
desire, and the greater desire brings the greater scorn
and pain because of the deprivation of the thing
desired. Therefore Epicurus, who pursues the most
tranquil life, says with respect to vulgar love:

Sed fugitare decet simulacra et pabula amoris
Abstergere sibi atque alio convertere mentem,
Nec servare sibi curam certumque dolorem:
Ulcus enim virescit et inveterascit alendo,
Inque dies gliscit furor atque aerumna gravescit,
Nec Veneris fructu caret is qui vitat amorem,
Sed potius quae sunt sine paena commoda sumit. [45]

CIC.   What does the *meridian of the heart* mean?
TANS.   The meridian of the heart refers to the highest
and most eminent part of the will which the strongest,
most direct and most luminous rays enflame. It means
that the affection in question is not as though, in its
initial movement, nor as though in its final repose,
but is in a point between the two, when its fervor is
most intense.

XIV.

CIC.   But what is the meaning of that arrow aglow with
flames at the iron point, around which a noose is

---

thrill in the presentiment of joy, and unite in a fertile union, as
they join the saliva of their mouths and press and breathe with their
tongues, it is all in vain. For they can glean nothing from the other,
and they cannot penetrate and be wholly absorbed body in body...

[44]   Eccl. 1:18.
[45]   LUCRETIUS *De rerum natura* iv. 1055-1066:

...But one must fly from love's image and nourishment and deny
oneself and divert the mind elsewhere and not become enslaved
to sorrow and inevitable pain. For an ulcer grows and festers with
nourishing, and, in time, the frenzy increases and burdens us with
calamity. And he who avoids this passion does not miss the delights
of Venus, but, instead, he reaps those profits which carry no burden
with them...

twisted, and of the motto, *Amor instat ut instans*?[46]
How do you understand it?

TANS.   I would say it means that love never leaves him,
and eternally afflicts him with invariable pain.

CIC.   I well understand the noose, arrow and the flame
and I understand the words, *Amor instat*, but I cannot understand what follows: that love persists because it is both of one instant and is also insistent.
This lacks as much sense as if one would say,—he
has imagined this emblem as he has imagined it, carries it as he carries it; I understand it as I understand
it; it is worth what it is worth; or, I esteem it as I
esteem it—.

TANS.   The less one considers the more easily is he apt
to judge quickly and condemn. *Instans* is not to be
taken as the adjective which comes from the verb
instare. It is to be understood as a substantive which
means an instant of time.

CIC.   Then what does he wish to express when he says
that love persists as the instant persists?

TANS.   What does Aristotle mean in his book on Time,[47]
when he says that eternity is an instant and the whole
of time is nothing but an instant?

CIC.   How can this be, if there is no time so brief that
does not have many instants? Would he mean to imply
that a single instant encompasses the deluge, the Trojan war and this very hour of our life? I would like
to know how this instant can be divided into so many
centuries and years. I would also like to know why
we could not affirm by a similar measurement that
the line is no more than a point?

TANS.   As time is one and yet is divided into diverse temporal subjects, so the instant is one in all the diverse
parts of time. As I am the same one who was, who
exists now and who will exist in the future, so am I
the same person here at home, in church, in the fields
and everywhere.

CIC.   But why would you have the instant to be the
whole of time?

TANS.   Because if there were not the instant, there would
not be time, for time in essence and substance is nothing more but an instant. And this will suffice—if

---

[46] "Love persists as does the instant."
[47] ARISTOTLE *Physics* iv. 217b, 224a.

you have the wherewithal to grasp it (for I have no time to give you a pedantic discourse on the fourth book of the Physics)—to make you understand that he means that love attends him by a presence which lasts for no less than the whole of time; for the word *instans* here is not to be taken to mean a mere atom of time.

CIC.   This meaning ought to be specified one way or another, if we wish to avoid the motto's being viciously equivocal. Thus we ought to be free to understand him to mean either that his love is the love of one instant, that is, of one atom of time and of no consequence, or, on the contrary, as you interpret it, that his love is eternal.

TANS.   Indeed if these two contrary senses had been implied, the motto would be a farce. But it is not a farce, if you consider it well; for it is impossible that love in one instant, if instant means a point or an atom of time, should persist with him forever; it is necessary, therefore, to understand the instant in another sense. In order to end this debate, let us read the verse:

> One time it expands, another time it re-
> assembles; one time it builds, another time it
> destroys; one time it weeps, at another it
> laughs; one time it is sad, at another it re-
> poses; one time it stands upright, at another,
> it sinks down.

> One time it lends a hand, another, it with-
> draws itself; one time it moves us on, another,
> stops us; one time it brings life, another, death.
> Through all the years, months, days and hours
> love is present, strikes, burns and binds me.

> Continually it shatters me, ever destroys
> me and keeps me in tears. It is my doleful
> languor in each and every hour.

> It forever harasses and uplifts me, and is
> too powerful in dispoiling me. There is no in-
> stant when it does not harass me, no instant
> when it does not bring me death.

CIC.   I have understood the meaning perfectly; and I confess that everything corresponds very well. But I think it time to proceed to the next one.

XV.

TANS.　Here you see a serpent languishing in the snow where a laborer has thrown it, a nude boy burning in the midst of flames, and some other details and circumstances, all accompanied by the motto, *Idem, itidem, non idem.* [48] This emblem seems to me more enigmatic than the one before it. Thus I shall not flatter myself that I can give a perfect explanation of it. However, I should think it meant that the same molesting fate torments both the boy and the serpent in a similar way (with intensity, without mercy and to the point of death) by those diverse and contrary principles of heat and cold. But I believe this requires longer and more　detailed consideration.

CIC.　Once again, read the verse.

TANS.

Languid serpent, you writhe, shrink, rise
and sink in that dense humour; and to ease your
intense pain, you move from one part of the
cold to another.

If the ice had ears to hear you, you a
voice to speak or to reply, I believe you would
have an efficacious argument to render it merciful to your torment.

I am tossed, consumed, burned, scorched in
the eternal fire, and in the ice of my goddess
neither love of me nor pity finds any place for
my delivery. Ah me, because she does not feel
how great is the rigor of my ardent flame!

Snake, you seek to escape, but you are
powerless. You cling to your shelter, but it
is dissolved. You call back your own forces,
but they are spent. Your hope is turned to the
sun, but a dense mist conceals it.

---

[48] "The same, in the same way, yet not the same." De Bèze employs an emblem of a serpent whose entrails are devoured by its offspring. The serpent is the Church. The tiny snakes are the Roman apostates. However, in accordance with his Calvinistic purpose, De Bèze's verse tells us this Church will live on while its apostates will be destroyed. See DE BEZE, *emblème* 32, p. 272.

You ask mercy of the laborer, and he hates
your sting. You invoke fortune, but senseless,
she does not hear you. Neither flight, refuge,
force, the stars, man nor fate can save you from
death.

You are hardened by the cold, while I am
liquefied by the heat; I wonder at your rigor,
you wonder at my ardor; you lust after the evil
I suffer, and I, after your desire.

Neither can I relieve your distress, nor
can you relieve mine. Now, aware enough of our
cruel fate, let us abandon all hope.

CIC.   Let us go now, so that as we walk we shall find a
way to untie this knot, if possible.

TANS.   Good.

END OF THE FIFTH DIALOGUE

AND THE FIRST PART OF

THE HEROIC FRENZIES

SECOND PART OF THE HEROIC FRENZIES

FIRST DIALOGUE

INTERLOCUTORS

CESARINO   MARICONDO [1]

I.

CES. They say the best and most noble things in the world take place when the entire universe is in the most perfect harmony with respect to all its parts. And this harmony is believed to occur when all the planets under the sign of Aries in the eighth sphere reach out to become a part of the invisible and superior firmament where the other zodiac is. They maintain that the worst and the most vile things take place when an inverse order and a contrary disposition predominates. Moreover, because of a vicissitudinal force, extreme mutations of things are known to take place between similar and dissimilar and between one contrary disposition and the other. Therefore, the revolution and the great year of the world [2] is that space

---

[1] Francesco Maricondo according to the *Fuochi di Nola* in the archives of Naples belonged to a noble family of Nola. He died in 1563 and left three infants. It is possible he is the interlocutor whom Bruno employs in this dialogue.

The name of Cesarino is also found frequently in the *Fuochi* and probably refers to Gian Domenico Cesarino, a soldier and comrade in arms of Gioan Bruno, born 1521. See *Fuochi di Nola*, fol. 67 r°, foc. 594 and fol. 230 v°, foc. 1877. See also VINCENZO SPAMPANATO, *Vita*, p. 37 and note 1, and BRUNO, in *Opere*, Gentile, II, 429 and note 1.

[2] Bruno held "the great year" was 36,000 years of revolution and was about to be completed. The fulfillment of the cosmic year had been pre-

of time during which there is a return to a certain
state of things, after others, definitely varied and opposite, have been traversed; as among the particular
years we see in the one called the solar year, that the
beginning of one contrary season is the end of the
other, and the end of that other is the beginning of
a new season. This is why we who today are in the
lowest ebb of the sciences, which have bred the scum
of opinions, themselves the causes of the vilest habits
and works, can certainly expect the return to better
conditions.

MAR.   Certainly this succession and order of things is
most true, my friend. However, as for ourselves, whatever may be our circumstances, the present afflicts
us more than the past does, and both present and past
together please us less than the future can; for we
always hold the future in expectation and hope, as
you can see very well represented by this emblem
borrowed from ancient Egypt. The Egyptians have
left us a particular statue in which three heads rose
from the same bust; one of a wolf who looked behind
him, the other of a lion who looked to one side, and
the third of a dog who looked ahead, in order to indicate that things of the past afflict us by the memory
of them, but not as much as things of the present
torment us in fact, while the future always promises
better things. Accordingly this emblem contains a wolf
who howls, a lion who roars and a dog who laughs. [3]

CES.   What does the motto written above it express?

MAR.   Notice that over the wolf is the word, *Iam;* over
the lion, *Modo,* and over the dog, *Praeterea,* words
which represent the three parts of time.

---

dicted in 1584 by the Bohemian astrologer, Cyprien Léowicz. The work of
Leowicz, *De conjunctionibus magnis insignioribus superiorum planetarum,*
originally published at Leuingen, Bavaria in 1564 was reprinted in London
in 1573 by Thomas Vautrollier, and was apparently available to Bruno.
See also BRUNO, *Des fureurs héroïques,* ed. Paul-Henri Michel, p. 354 and
note 1.
    [3] See PIERO VALERIANO, *Hieroglyphica* (Francfort, 1678), pp. 192, 384,
for the view that the Egyptian triple-head of wolf, lion and dog reflect
the three modes of time, past, present and future. An early edition had
appeared in 1556. Valeriano's treatise was based on Horapollo's *Hieroglyph-
ica,* 1419, which influenced countless emblem books of the sixteenth century.
See ERWIN PANOFSKY, *Meaning in the Visual Arts* (Anchor, 1955), pp. 149,
151-152, 158-162.

CES.   Now read what is written on the tablet.
MAR.   I intend to do precisely that.

A wolf, a lion and a dog—at dawn, in the bright-
[ness
of day and in the dark of evening—represent the
things I have spent, the things I retain and the things
I shall gain of all that has been given me, is given to
me and can be given to me.

For the things I have done, do now and must do, in
the past, present and in the future, I repent, am tor-
[mented
and am assured, in regret, in suffering and in expec-
[tation

The harshness of my past experience, the bitterness
[of its
fruit and the sweetness of hope are a menace, an affliction and a solace to me.

The years I have lived, the time I live now and
[shall
live,—the past, present and future—make me
tremble, excite me and sustain me.

What has gone by, what happens now and what
[will
follow, holds me in much fear, in too much martyr-
[dom, and
yields me sufficient hope.

CES.   This is precisely the head of a frenzied lover; and
very likely of all mortals who are afflicted, whatever
may be the manner or mode of their affliction; for
we cannot say, nor ought we to say that such a destiny corresponds to all in general, but only to those
destinies which were or are laborious. For example, it
behooves one who has sought a kingdom and now
possesses it to feel the fear of losing it; it behooves one
who has labored to acquire the fruits of love and to
know the special favor of the beloved to feel the bite
of jealousy and suspicion. And with respect to our condition in this world, if we find ourselves in darkness
and misfortune, we can safely prophecy light and pros-
perity; if we live in an era of felicity and enlighten-

ment, without doubt we can expect a succession of affliction and ignorance. [4] For example, Mercury Trismigistus saw Egypt in such a great splendor of science and of prophetic wisdom that he esteemed men to be the brothers of both demons and gods, and consequently to be most inspired; nevertheless to Asclepius he made that prophetic lamentation which announced that there must follow a dark age of new religions and cults, and that Egypt's present splendor would become only a fable and a matter for condemnation. [5] Similarly, when the Hebrews were slaves of Egypt and exiled in the desert, they were comforted by their prophets who assured them of liberty and the conquest of a fatherland, but when they enjoyed a state of power and tranquillity, they were menaced by captivity and dispersion. And today there is no evil or dishonor to which we may be subject, that we may not expect honor and goodness tomorrow. The same befalls other generations and states. If these states endure and are not ever annihilated, they must pass from evil to good, from good to evil, from baseness to splendor, from splendor to obscurity by a necessary force of the mutations of things. For this vicissitude occurs in accordance with the natural order. And if one should find another order which would alter or correct the present one, then I would consent to it, and would have no way in which to dispute it, for I judge only by the light of my natural reason.

MAR.   We know that you are not a theologian but a philosopher, and that you treat of philosophy, not of theology.

CES.   That is the case. But let us see what follows.

II.

CES.   Next I see an arm upholding a smoking incense burner, bearing the motto, *Illius aram;* [6] and following the emblem is the sonnet:

---

[4] Bruno employs the concept of the three modes of time, past, present and future in this passage to suggest the historical theory of cyclic recurrence.

[5] For a full statement of Trismigistus' prophecy see BRUNO, *Spaccio della bestia trionfante,* in *Opere,* Gentile, II, 192-193 and note 1.

[6] "His altar." For Alciati's emblem of a hand grasping a taper see Alciati, plate 38. For John Donne's allusion to lovers as "tapers" see "The Canonization", st. 3. For William Shakespeare's similar allusion to lovers as "tapers" with their ladies' "sweet breath puffed out" see *Love's Labour's Lost,* V, ii, 267-270.

Who would deem that transport of my lofty
passion less worthy of the divinity, because it
is expressed in the painted flourish of my vows
on tablets offered in the temple of fame?

Though I am called to another and more
heroic enterprise who will ever deem it less
becoming for this beauty to hold me captive
of its external worship, when heaven itself
so loves and honors it?

Leave me, leave me, other desires, importunate thoughts, leave me in peace! Why
do you wish me to withdraw

from the sight of the sun that delights me
so? But you, oh my thoughts, filled with pity,
say to me:—Why do you contemplate an object
whose contemplation consumes you?

Why are you so smitten by that flame? I
reply: Because this torment contents me more
than any other pleasure.

MAR. With respect to this verse I tell you that, no matter how much one remains attached to corporeal
beauty and to external veneration of it, he may still
conduct himself honorably and worthily; for from
material beauty, which reflects the splendor of the
spiritual form and act and is its vestige and shadow,
he will arrive at the contemplation and worship of
divine beauty, light and majesty. Thus from visible
things he begins to exalt his heart toward those things
which are so much the more excellent in themselves
and pleasing to the purged soul, because they are
more removed from matter and sense. Oh God, he
will say, if a shadowy, cloudy, elusive beauty painted
upon the surfaces of corporeal matter pleases me so
much and so incites my passion, so influences my spirit
with I know not what reverence of majesty, so capti-
vates me and so sweetly binds me and draws me to
it, that I find my senses offer nothing so agreeable to
me, what would be the effect upon me of that which

is the substantial, original and primal beauty? What
would be the effect of that beauty upon my soul, upon
a divine intellect and upon the order of nature? Therefore, the contemplation of this vestige of light must
lead me by the purgation of my soul to a resemblance,
a conformity and a participation in that most worthy
and most lofty light into which I am transformed and
to which I am united. For I am sure that nature, having put this (corporeal) [7] beauty before my eyes and
having endowed me with an interior sense through
which I can discern the most profound and incomparably superior beauty, wishes that from here below
I become elevated to the height and eminence of
that most excellent species. [8] Nor do I believe that my
true divinity, inasmuch as it is shown to me in its
vestige and image, would be offended if I happened
to honor it in its vestige and image and to offer sacrifices to it, provided the impulse of my heart remained, ordered and my affection remained intent
upon the higher good; for who is that man who can
honor the divinity in its essence and its own substance,
if in its essence and substance he is unable to comprehend it?

CES.   You have demonstrated quite well how men of heroic spirit convert everything to good and how from
captivity they know how to nurture the fruits of a
greater liberty, and in the experience of defeat how
to find the occasion of the greatest victory. You know
very well that to men who are well disposed the love
of material beauty not only does not at all delay them
from the greater enterprises, but rather gives them
wings to accomplish them; for love's constraint is
transformed into a virtuous zeal which forces the
lover to progress to the point of becoming worthy of
the thing loved, and perhaps worthy of some greater
and still more beautiful object; so that either he begins to feel content that he has gained his desire, or
he is gratified that the particular beauty of his object
gives him just reason to scorn any other as a beauty
that he has conquered and surpassed; consequently,

---

[7] Parentheses mine.
[8] PLOTINUS *Enneads* 1. 6. 2.

either he rests in tranquillity, or bestirs himself to aspire to more excellent and more magnificent objects. For this reason the heroic spirit constantly renews its efforts, as long as it does not see itself uplifted toward the desire of the divine beauty in itself, that is, the beauty without similitude, analogy, image or species, if such a beauty were possible; and if it were possible for the heroic spirit to know how to attain it.

MAR. You see then, Cesarino, how this frenzied one is right in resenting those who reprove him as captive of a base beauty to which he offers vows and tablets. [9] For his captivity does not make him a rebel against the voices which call him to the higher beauties, inasmuch as ignoble objects derive from lofty objects and are dependent upon them, and it is from these base objects that he is able to have access to these higher objects in due degree. Those objects, if not God, are things divine and are living images of God, and he is not in the least offended at seeing himself adored in them, for have we not the command of the supernal spirit who says, *Adorate scabellum pedum eius?* [10] And elsewhere has not the divine ambassador said, *Adorabimus ubi steterunt pedes eius?* [11]

CES. God, the divine beauty and splendor, shines and is in all things; but to me it does not seem erroneous to admire him in all things according to his mode of communication. What would certainly be erroneous would be to give others the honor due to him alone. But what does he mean when he says, *Leave me, leave me, other desires?*

MAR. He banishes certain thoughts from himself, because they present him with other objects which, though not having any power to move him, yet would steal from him the view of the sun, a view he can see through this window [12] more than through any other.

CES. Why, troubled by these thoughts, does he remain constant in gazing on that splendor which ruins him

---

[9] i., e., upon the altar in the temple.
[10] Ps. 98:5: ...Exalt ye the Lord our God, and adore his footstool, for it is holy...
[11] *Ibid.*, 131:7: ...We will go into his tabernacle; we will adore in the place where his foot stood...
[12] i., e., of his particular object.

and does not give him any pleasure unaccompanied with severe torment at the same time?

MAR.   Because in this discordant life all our consolations are accompanied by discomforts which are equally abundant. For example, the fear of a king in the peril of losing his kingdom is greater than the fear of a beggar who risks the loss of ten farthings; the solicitude of a prince for his republic is more urgent than the care of a shepherd for his flock of sheep; but the pleasures and delights of the king and the prince are perhaps greater than the pleasures and delights of the shepherd. Therefore to love and aspire higher is accompanied by the greater glory and majesty, but is also accompanied by the greater care, sadness and pain. I mean that in our present state where one contrary is always joined to the other, the greatest contrariety is always found in the same genus, and, consequently, with respect to the same matter, even though these contraries may not exist simultaneously. And similarly, in proportion one can apply to the love of superior Cupid those things which the Epicurean poet affirms of vulgar and animal love when he says,

Fluctuat incertis erroribus ardor amantum,
Nec constat quid primum oculis manibusque fruantur:
Quod petiere, premunt arte, faciuntque dolorem
Corporis, et dentes inlidunt saepe labellis
Osculaque adfigunt, quia non est pura voluptas
Et stimuli subsunt qui instigant laedere id ipsum,
Quodcunque est, rabies, nude illa haec germina sur-
                                        [gunt.
Sed leviter paenas frangit Venus inter amorem,
Blandaque refraenat morsus admixta voluptas;
Namque in eo spes est, unde est ardoris origo,
Restingui quoque posse ab eodem corpore flam-
                                        [mam. [13]

---

[13] LUCRETIUS *De rerum natura* iv. 1077-1087:

...The passions of lovers fluctuate in wavering uncertainty and they cannot agree what things to enjoy with their eyes and hands. For as they seek their joy they press the object of love so tightly that they bring pain to-the body. And they kiss so hard that their teeth drive into their lips, because their desire is not unmixed. They are goaded on by an instinct to injure whatever sprouts forth from this germinating madness. But in love Venus lightens the

It is by these enticements, then, that nature's power and skill cause one to be consumed by the pleasure of what destroys him, bringing him content in the midst of torment and torment in the midst of every contentment, for nothing results from an absolutely uncontested principle, but everything results from contrary principles through the triumph and conquest of one of the contraries. There is no pleasure of generation on the one hand without the displeasure of corruption on the other; and where things which are generated and destroyed are found to be conjoined and as though composed in the same subject the feeling of delight and sadness is found at the same time; but more readily is it called delectation rather than sadness, if it happens that delectation predominates and solicits the sensibility of the subject with greater impact.

III.

CES.   Now let us contemplate the emblem of a phoenix burning in the sun. By its smoke the phoenix almost obscures the splendor of the sun whose fire inflames it; and there is a motto which says, *Neque simile, nec par.* [14]

MAR.   Let us read the verse first:

This phoenix which kindles itself in the golden sun and bit by bit is consumed, while it is surrounded by splendor, returns a contrary tribute to its star;

because that which ascends from it to the sky, becomes tepid smoke and purple fog, which cause the sun's rays to remain hidden from our eyes, and obscures that by which it glows and shines.

Thus my spirit (which the divine splendor inflames and illumines), while it goes about

penalties she imposes, and moderates the anguish by blending pleasure with pain; for in love there is the hope that the flame of passion may be quenched by the same body that fanned it...

[14] "Neither similar nor equal to it." See above, p. 154 and note 20.

explaining that which glows so brightly in its thoughts,

sends forth verses from its high conceit,
only to obscure the shining sun, while I am
completely consumed and dissolved by the effort.

Ah me! This purple and black cloud of
smoke darkens by its style what it would exalt,
and renders it humble. [15]

CES.   This verse tells us then, that as the phoenix, set
on fire by the splendor of the sun and accustomed to
its light and flame, sends forth to the sky smoke
which obscures the very sun that kindled it, so the
frenzied one inflamed and illumined by his every ef-

---

[15] DANTE *Convivio* 3, 2:

Amor che ne la mente mi ragiona
de la mia donna disiosamente,
move cose di lei meco sovente,
che lo 'ntelletto sovr'esse disvia.
... ...
Però, se le mie rime avran difetto
ch'entreran ne la loda di costei,
di ciò si biasmi il debole intelletto
... ...
Cose appariscon ne lo suo aspetto,
che mostran de' piacer di Paradiso,
... ...
Elle soverchian lo nostro intelletto,
come raggio di sole un frale viso
e perch' io non le posso mirar fiso,
mi convien contentar di dirne poco.
... ...

Love that discourses to me in my mind, yearningly
of my lady moves many a time such things with
me about her that my intellect loses its way
concerning them.
... ...
Therefore if defect shall mark my verses which
shall enter upon her praises, let our feeble
intellect be blamed...
... ...
Things are revealed in her aspect which show us of
the joys of Paradise. ...They transcend our
intellect, as the sun's ray the feeble vision;
and because I may not fixedly gaze upon them,
I must content myself with scant speech of them...

fort to offer praises to the brilliant object that has enkindled his heart and enlightened his thought, succeeds more in obscuring the object than in giving it any of his own light; for like the phoenix, he sends up smoke caused by the flames in which his substance is dissolved.

MAR. Without wishing to weigh and compare the labors of this lover, I return to what I was telling you the other day, that praise is one of the greatest sacrifices human passion can offer to its beloved object. And, putting aside matters which touch of the divine, tell me this. Who would know about Achilles, Ulysses and so many other Greek and Trojan captains, who would guard the memory of so many great soldiers, men of wisdom and heroes of this world, if they had not been raised to the stars and deified by the sacrifice of praise upon an altar enkindled in the hearts of poets and other illustrious seers, a sacrifice which raises to the sky the celebrant, the victim and the divine hero, canonized by the hand and vow of a legitimate and worthy priest?

CES. You do well to say *a worthy and legitimate* priest, for there are many false priests in the world today, who, themselves unworthy, usually celebrate others who are as unworthy as they are, just as *asini asinos fricant.* [16] But according to the will of Providence, instead of both ascending to heaven, both will descend jointly into the darkness of Orcus; so that the glory which both the celebrant and the celebrated receive will be vain, for one has interwoven a statue of straw, or cut a trunk of wood, or cast a piece of cement; and the other, an idol of infamy and baseness, fails to realize that he will not have to wait for the bite of old age or the scythe of Saturn to cut him down, for he will be buried alive by his own panegyrist in the same hour of the eulogy that hails, elects and exhibits him. A contrary recompense fell to the prudence of that most celebrated Maecenas. If this man had not had any other renown than a spirit inclined to the protection and favor of the Muses, that renown alone would have merited him the respect of so many illustrious poets whose genius set him among the most famous heroes who have walked the face of the earth.

[16] ...jackasses mock jackasses...

His own studies and his own renown rendered him illustrious and most noble, and not his birth from a race of kings, nor his position as chief secretary and counselor of Augustus. What has made him most illustrious, I say, is to have rendered himself worthy of the fulfillment of the promise of that poet who said.

Fortunati ambo, si quid mea carmina possunt,
Nulla dies nunquam memori vos eximet aevo,
Dum domus Aenae Capitoli immobile saxum,
Accolet, imperiumque pater Romanus habebit. [17]

MAR.   I am reminded of what Seneca says in a certain epistle in which he refers one of his friends to the following words of Epicurus: "If it is love of glory that moves your heart, my letters will render you more noteworthy and illustrious than all these other things you honor and which give you honor, and of which you may boast. [18] Homer might have been able to say the same thing to Achilles, or to Ulysses if he could have faced them, and Virgil, the same thing to Aeneas and all his progeny. Therefore, as that moral philosopher well expressed it, "Idomeneus is better known because of the letters of Epicurus than are all the lords, satraps and kings upon whom his title depended, for the memory of those kings is obliterated in the deep darkness of oblivion. Atticus is known not because he was the son-in-law of Agrippa and the progenitor of Tiberius, but because of the letters of Tullius. Drusus, the great-grand-nephew of Caesar, would not be among the number of great men if Cicero had not placed him there. Indeed, the high flood of time submerges us, and above that flood few men of genius will raise their heads". [19]

Now let us return to the argument of this frenzied one who, seeing a phoenix burning in the sun, is re-

---

[17] VIRGIL *Aeneid* ix, 446-449:

...Both of us are fortunate, for if my verse can mean anything, no length of days shall ever blot you from the memory of time, while the house of Aeneas shall dwell by the steadfast Capitolian rock, and the Roman lord hold sovereignty...

[18] SENECA *Epistolae* 21. 3.

[19] *Ibid.*, 21. 4, 5.

minded of his own zeal and laments that like the phoenix he returns the light and fire he receives in nothing but an obscure and tepid smoke of praise in the holocaust of his own dissolving substance. As a result, we can never make divine things the subject of our thought without detracting from them rather than adding any glory to them, so that the best thing a man can do with respect to them is to seek rather to ennoble himself in the presence of other men by his own zeal and ardor than to give praise to another by some complete and perfect act. For such an act cannot hope to make progress toward the infinite in which unity and infinity are one and the same, in the pursuit of which one vainly binds himself to any other kind of number; for the infinite is not a unit or any kind of unit, because it is not a number, or any unit of numbers, for no number or unit of numbers can be the same thing as the absolute or the infinite. Accordingly a theologian says well that, inasmuch as the fount of light not only far exceeds our finite intellect but also exceeds divine ones, it is proper to celebrate it not with speeches and discourses, but in silence. [20]

CES.   Yes. But not with the silence of brute animals and those who have but the image and likeness of men, but with the silence of those whose silence is more illustrious than all the screeches, noises and uproars that can be heard.

IV.

MAR.   But let us continue and see what the other emblems mean.

CES.   Tell me if you have already seen and considered the meaning of this fire in the form of a heart with four wings, two of which have eyes. The entire figure is encircled by luminous rays and by the inscription, *Nitimur in cassum?* [21]

MAR.   I recall well that this must represent the state of mind, heart, spirit and eyes of the frenzied one; but let us read the sonnet:

---

[20] DIONYSIUS THE AREOPAGITE, *Liber de Trinitate*, ed. Ficino (Bale, 1561), p. 1021.

[21] "Are we searching fruitlessly?"

As these thoughts aspire to the holy
splendor, no sublime effort delivers them of
obscurity; and the heart which those thoughts
would refresh is unable to withdraw itself
from woe.

The spirit, which would welcome a brief
truce, is denied one moment of pleasure; and
eyes that would be closed in sleep all the
night long are wide with weeping.

Ah me, eyes of mine, by what labor and
art can I calm my afflicted senses? Spirit
of mine, when and where

shall I temper your intense pain? And
you, heart of mine, how shall I offer you the
appeasement to compensate for your grave
torment?

When will the soul provide you with your
due, oh afflicted mind whose heart, spirit and
eyes share your complaint?

Because the mind aspires to the divine splendor it
flees association with the crowd and withdraws itself
from the multitudes, but it also flees their pursuits,
judgments and opinions; for there is the greater danger of contracting ignorance and vice the greater the
multitude with whom one becomes confounded. *In
public spectacles*, says a moral philosopher, *in the
midst of pleasure the more easy it is to engender
vices.* [22] If this man desires the highest splendor, he retires as much as he can to the one and withdraws
within himself as much as possible, so that he may not
be like the multitude of men who constitute the
majority; and he would not be their enemy because
they are different from him; but he gains the good will
of one and another of them if he can; otherwise he
interests himself in the one that seems better to him.
He converses with those whom he can make better,
or those who can make him better, by the light he
can give them, or the light they can give him. He is

---

[22] SENECA *Epistolae* 7. 2.

happier with one worthy individual than with an inept multitude. Nor does he believe he has achieved little when he has become wise in himself, for he remembers the words of Democritus, *Unus mihi pro populo est, et populus pro uno;*[23] and those words which Epicurus wrote to a fellow student, *Haec tibi, non multis; satis enim magnum alter alteri theatrum sumus.*[24]

The mind, then, which aspires to raise itself first turns from the multitude, considering that the light above us scorns our strife and is to be found only where the intelligence is, and not where every intelligence is but that one which, of those that are few, principal and first, is the unique, prime, principal and one.

CES.   How do you understand that the mind aspires to raise itself? For example, would it be by turning towards the stars, or the empyrean, or the crystalline heaven?

MAR.   Certainly not, but by proceeding to the depths of the mind; and in order to accomplish this, it is not at all necessary to gaze wide-eyed toward the sky, to raise one's hands, to direct one's steps toward the temple, wearying the ears of statues with the sounds we make; but it is necessary to descend more intimately within the self and to consider that God is near, that each one has Him with him and within himself more than he himself can be within himself, for God is the soul of souls, the life of all lives, the essence of essences; and the planets you see above and below the canopy of heaven (as it pleases you to call it), are only bodies, creations similar to our earth, in which the divinity is present neither more nor less than it is present in this body which is our earth as well as in our very selves.[25] These are the reasons why one

---

[23] SENECA *Epistolae* 7. 10.

...I prefer the one to the multitude, and so do the people...

[24] SENECA *Epistolae* 7. 11:

...These things belong to you, not to the many; indeed we are a sufficiently magnificent mirror to each other...

[25] BRUNO, *La Cena de le Ceneri,* in *Opere italiane,* I, 27:

...Cossí siamo promossí a scuoprire l'infinito effetto dell'infinita causa, il vero e vivo vestigio de l'infinito vigore; ed abbiamo

must first of all leave the multitude and withdraw within himself. Then he must reach the state in which he no longer regards but scorns each struggle, so that the more passion and vice fight him from within and vicious enemies from without the more will he recover his breath and rise again, and with one exhalation (if possible) surmount the steep ascent. Then he is in no need of arms and shields other than the greatness of an unconquered soul and the endurance of spirit capable of maintaining his life in equilibrium and continuity, a spirit which proceeds from knowledge and is regulated by the art of speculating upon things lofty as well as base, upon things divine as well as human; and it is in this speculation that the highest good consists. Consequently a moral philosopher said, writing to Lucilius, that it was not necessary to pass through the straits of Scylla and Charybdis, or to penetrate the deserts of Candavia and the Appenines, or to leave the Syrtes behind; for our path is as secure and pleasant as nature herself could arrange. And he said that it is not gold or silver which makes man similar to God, because God does not amass such treasures; it is not adornments, for God is naked; it is not ostentation or fame, for very few are those to whom he exhibits himself, and perhaps no one knows him, and, indeed, many and more than many have a false idea of him; neither is it the possession of so many things we ordinarily admire, for it is not the desire for the abundance of these things that makes us rich, but our contempt for them. [26]

CES.   Good. But tell me now in what way our poet will *calm his senses, and temper his spirit's pain, appease his heart and give his mind its due,* so that in his aspiration and zeal he shall not have to ask, *are we searching fruitlessly?*

---

dottrina di non cercare la divinità rimossa da noi, se l'abbiamo appresso, anzi di dentro, piú che noi medesimi siamo dentro a noi...

...So we advance to discover the infinite effect of an infinite cause, the true and living shadow of the infinite force; and we have the wherewithal not to search for a divinity removed from us, since we have it near and, in fact, within us, more than we ourselves are within ourselves...

[26] SENECA *Epistolae* 31. 9, 10.

MAR.   He may accomplish all these things by realizing that his soul is in his body in such a way that its superior part may be removed to join and attach itself to divine things as by an indissoluble vow. In that state he will feel neither hate nor love of mortal things, for he will prefer to be the master rather than the servant and slave of a body he regards as nothing more than a prison which holds his liberty in chains, a snare which entangles his wings, a chain which holds fast his hands, shackles which have fixed his feet, and a veil which obscures his vision. But at the same time he will not feel himself a servant, captive, ensnared, enchained, impotent, impenetrable and blind, because his body will not tyrannize over him any more than he himself allows it to, for now his body will be subjected to his spirit in the same way that matter and the corporeal world are subject to the divinity and to nature. Therefore, he will render himself strong against fortune, magnanimous before injuries, dauntless against poverty, diseases, and persecutions.

CES.   Then this heroic frenzy is well integrated.

V.

CES.   Let us look at the following emblem which depicts a wheel of time moving about its own center, with the motto, *Manens moveor.* [27] How do you understand this?

MAR.   It means that the wheel turns upon itself, so that motion and rest concur, for the spherical motion of a body upon its own axis and its own center implies the rest and immobility associated with rectilinear motion; or, one may say, there is a certain repose of the whole and a motion of its parts; and the parts which are moved in a circle have two kinds of alternate movement, inasmuch as some parts ascend to the summit, while others in turn descend to the bottom; some parts remain in an intermediate position, and some remain in the extreme position either at the top or bottom. And it appears that all this has to do with the subject of the following sonnet:

---

[27] "While standing fixed, I am moved." See above, introduction, p. 53.

That which my heart holds both clear and obscure, beauty engraves in me but humility erases. Zeal sustains me, but another care brings me to the source of all the labors of my soul.

When I think of tearing myself away from the pain, hope revives me, (while) [28] the vigor of another thought binds me; while love raises me, reverence debases me as I aspire to the noblest and the highest good.

Lofty thought, holy desire and intense zeal of mind, heart and labor, to the immortal, divine but immense object

join me, enwrap me in it, and cause it to nourish me. No longer may my mind, reason and sense strive elsewhere, discourse, or become elsewhere entangled.

So that one may say of me: This one who has now fixed his eyes upon the sun, and, become a rival of Endymion, is grieved.

Therefore, the continual motion of the one part of the wheel supposes and leads with it the motion of the whole, and the hurling down of the upper parts causes a drawing up of the lower parts; thus, the impulsion given by the superior parts necessarily results in the inducement of the inferior ones, and from the descent of a potency follows the ascent of the opposite potency. At this point the heart (which represents all the affections in general) becomes obscure and translucent, restrained by its zeal, raised by magnificent thoughts, reinforced with hope, weakened by fear. And in this state and condition those who find themselves subject to the destiny of generation will ever be seen.

VI.

CES.   Very good. Let us pass to the emblem which follows. I see a ship inclined upon the sea; its ropes

---

[28] Parentheses mine.

are attached to the shore and it bears the motto, *Fluctuat in portu.* [29] Explain what it can mean; if you have resolved the enigma, enlighten me.

MAR. The emblem and the motto have a certain kinship with the preceding emblem and motto, as can be seen easily, if one reflects a little. But let us read the sonnet:

If heroes, gods and men encourage me not
to despair, no fear of death, no pain of the body,
no impediments of pleasure

will cause me excess terror, suffering or desire;
and that I may clearly see my path before me, may
doubt, pain and sadness be extinguished by hope,
joy and inner delight.

But if the being who now renders my thoughts
so uncertain, my desires so ardent and my pleas so
vain, should deign to look upon those thoughts, fulfill
those desires and listen to those pleas,

no one who dwells in the abode of birth, life and
death would be capable of such happy thoughts,
[accomplished
desires and pleas granted;

when heaven, earth and hell stand in the way, if my
divinity shine upon me, enkindle me and hold me near,
she will give me light, power and beatitude.

We understand the sentiment expressed here in the light of our explanation in the preceding discourses, expecially where we have shown that the sense of inferior things is attenuated and even nullified when the superior powers are valiantly intent upon the more glorious and heroic object. So great is the virtue of contemplation (as Iamblicus notes) that sometimes the soul not only turns itself from inferior acts, but also

---

[29] "It floats in port." Alciati's emblem of a ship at sea with two men leaning lackadaisically over the rail indicates it is easy to withdraw from virtue. See Alciati, plate 53. Variant symbols of a ship at sea occur frequently in alchemical manuscripts and symbolize the soul's arduous search for fullness of being. See MICHAEL MAIER, *Viatorum* (Rouen, 1651), p. 183.

escapes the body completely. I would understand this only according to the several modes enumerated in the book of *The Thirty Seals*. [30] This book presents all the varieties of contraction, by which some ignominiously and others heroically arrive at the point of no longer feeling the fear of death, or suffering the pain of the body, or feeling the impediments of pleasure; for hope, joy and the delights of the higher spirit gather such force, that they abolish all the passions which can engender doubt, pain and sadness.

CES.   But whom does the lover summon to look upon his thoughts rendered so uncertain; whom does he ask to fulfill those desires which have become so ardent; and whom does he ask to listen to those pleas rendered so vain?

MAR.   He addresses the object which gazes upon him the moment he shows himself to it; for to see the divinity is to be seen by it, just as to see the sun is to be seen by it. In like manner, to be heard by the divinity is precisely to hear it, and to be favored by it is the same as to offer oneself to it. For the divinity is one, immovable and always the same, from whom proceeds those uncertain and certain thoughts, tormenting and pleasing desires, pleas which are refused and granted, accordingly as man unworthily or worthily presents himself to it with his intellect, affection and activity. Similarly, the pilot of a ship is called the occasion either of the sinking or the salvation of the ship, accordingly as he stays with it, or is found to have abandoned it. However, it is by his delinquency or conscientiousness that the pilot ruins or saves the ship, while the divine power, which is all in all does not offer or withdraw itself except by the conversion or aversion of someone else.

VII.

MAR.   It seems to me, then, that there is a strong connection between this and the following emblem in which we find two stars in the shape of two radiant eyes and the motto, *Mors et vita*. [31]

---

[30] Published by Bruno in London in 1583 and referred to frequently as the *Art of Memory*.

[31] "Death and life." Alciati's emblem depicts a hand suspended in mid-air; the hand has a large opened eye in its palm. The hand is meant to guide man along the path of prudence. See Alciati, p. 22.

CES.   Read the sonnet then.
MAR.   I shall.

You can see written on my face by the
hand of love the history of my pain. But
because your pride knows no restraint and I am
eternally unhappy,

you allow your beautiful eyelids, so
cruel to me, to hide your delightful eyes, so
that the murky sky does not clear, and the
baneful and inimical shadows do not dissolve.

By your great beauty and by that love of
mine which almost equals it, render yourself
merciful, goddess, for love of God.

Do not prolong this too intense evil,
which is an undeserving penalty for my abundant
love. Let not too much austerity accompany
such splendor!

If you condescend that I may live, open
the gates to your gracious glance. Gaze upon
me, oh lovely one, if you wish to give me death. [32]

The face upon which the story of his pain is written
is the lover's soul, inasmuch as it is exposed to blessings
from on high; with respect to those blessings the
soul exists only in potency and aptitude without the
accomplishment of that perfect act which awaits the
divine dew. Thus was it well said, *Anima mea sicut
terra sine aqua tibi.* [33] And elsewhere, *Os meum aperui*

---

[32] GUIDO CAVALCANTI *Rime* 18:

> Morte gentil, remedio de' captivi,
> merzè, merzè a man giunto ti cheggio,
> vienmi a vedere e predimi...

> Gentle death, remedy of the enslaved,
> mercy, mercy with folded hands I beg you,
> come visit me and take me...

[33] Ps. 142:6:

> ...I stretched forth my hands to thee: my
> soul is as earth without water unto thee...

et attraxi spiritum, quia mandata tua desiderabam. [34]
Next that *pride which knows no restraint*, is a metaphorical allusion. For God is often called jealous,
angry or asleep and the metaphor indicates how difficult God makes it for us to see even his shoulders;
that is, to see him even by his vestiges and effects.
Thus he shuts out the light with his eyelids, and does
not bring calm again to the murky sky of the human
mind by removing from it the shadow of enigmas and
similitudes. Nevertheless (because he does not believe
that what has not yet happened will never happen),
the frenzied one begs that the beauty of the divine
light be not concealed from everyone, but at least show
itself according to the capacity of him who contemplates it. And he begs that beauty in the name of his
own love, which is perhaps equal to it (that is, equal
to that beauty inasmuch as he can make himself comprehensible to it), to be merciful to him, so that it
may make him like those others who are gentle and
who, from crude and distant become benign and affa-
ble. He entreats that beauty not to prolong the evil
that comes from being deprived of it, and asks it not
to allow the splendor he desires to please him more
than the love by which it can communicate with him;
for all the perfections found in the divinity are not
only equal one to the other, but are even one and the
same.

Finally he pleads again with the divinity not to
sadden him any longer by depriving him of itself; for
the divinity can bring him death by the light of its
eyes and by the same light can give him life; but if
it bring him death, he pleads that it be not by shut-
ting out the endearing light with its eyelids.

CES.   Does he refer to that death of lovers which proceeds from the supreme joy, called by the Cabalists
mors osculi, [35] the same thing as the eternity to which
man can be disposed in this life and realize fully here-
after?

MAR.   Precisely.

---

[34] Ps. 118:131:

> ...I opened my mouth and panted, because
> I longed for thy commandments...

[35] "The death of the kiss"; see also above, p. 127.

VIII.

MAR.   Now it is time to consider the next emblem which
is similar and related to the preceding ones we have
discussed. There is an eagle which flies up to heaven
on its two wings; but I do not know how much it finds
itself weighed down by a stone tied to one of its feet.
It's motto is, *Scinditur incertum.* [36] Without a doubt
the motto refers to the multitude, number and mass
of potencies of the soul; and that famous verse completes its meaning:

  Scinditur incertum studia in contraria vulgus. [37]

This multitude is generally divided into two fac-
tions (although when thus divided their powers are
not limited to two); thus, among the potencies of the
soul some incite us toward the loftiness of the intelligence and light of justice, while others lead, incite
and in a certain fashion force us to baseness, to the
filthiness of sensuality and to the satisfactions of natural instincts. Accordingly, the sonnet says:

> I long to do good, but it is denied me;
> my sun is not with me, although I am with it;
> for in order to be with it, I am no longer with
> myself, and the nearer I am to it, the further
> it is from me.

> For one moment of joy, I do much weeping;
> seeking happiness, I find affliction; because
> I look too high, I am blinded and to obtain my
> good, I lose myself.

> Through bitter sweetness and delightful
> pain, I fall to the center and am drawn up to-
> ward the sky; necessity constrains me while
> the good leads me on; fate draws me to the
> abyss, while counsel uplifts me; desire spurs
> me on, while fear bridles me; care burns me
> and keeps me long in peril.

---

[36] "Torn by uncertainty." Alciati's emblem shows a nude boy borne
upward by a hand that holds a dove, while his other hand carries the
weight of a heavy stone. See Alciati, plate 19.

[37] VIRGIL *Aeneid* ii. 39:

> The wavering crowd is torn apart by contrary disputes...

What straight or devious path will give
me peace, and free me from discord, if the
one rejects me so, and the other invites me? [38]

The ascent takes place in the soul by the vigor
and impulse of the wings which are the intellect and
the will. It is by these faculties that the soul naturally turns and fixes its gaze toward God as upon the
sovereign good and primary truth, the absolute goodness and beauty; just as every natural thing has a
regressive impulse toward its own origin, and a pro-
gressive impulse toward its own end and perfection,
as Empedocles had well explained, to whose opinion
I think the Nolan refers in the following octave:

It happens that the sun returns to its
point of departure, and its diffusive light
returns to its source; and what belongs to
the earth falls back to the earth; and the
rivers issuing from the sea flow again to the
sea, and desires aspire to the place from
which they have drawn their life and breath.
In the same way, born from my goddess, my
every thought to my goddess must return.

The intellectual faculty is never in repose, is
never pleased by any truth it attains, but proceeds
onward toward an incomprehensible truth. Similarly
we see that the will, which follows the cognition, is
never satisfied with anything finite. Therefore we conclude that it is the soul's nature to know no other end
than the origin of its substance and its entity. But
because of the natural potencies that dispose it to
the care and government of matter, the soul begins
to direct its impulse to serve and communicate its
perfection to inferior things, thus bearing witness to
its resemblance to the divinity, which communicates
itself by its goodness and either produces in an infinite way by giving being to an infinite universe and
the innumerable worlds within it, or in a finite way by
producing only this universe subject to our eyes and
to our mortal reason. Granted that it belongs to the
unique essence of the soul to have two kinds of po-

---

[38] See above, pp. 96-97.

tencies which order it towards its own and toward the lesser good, it is customary to depict it by a pair of wings, whose power impels it toward the object of its prime and immaterial potencies; and by a stone, whose weight re-establishes the aptitude and efficacy it has toward the objects of its secondary and material potencies.[39] That is why the inner affection of the frenzied one is amphibious, divided, afflicted, and more easily inclined toward the base than urged toward the higher things; for the soul, though exiled in an inferior and hostile land where its powers are enfeebled, partially inhabits a region far from its natural abode.

CES.   Do you believe this difficulty can be overcome?

MAR.   Very well. In the beginning the effort is most trying, but it becomes easier and easier as the progress of contemplation becomes more fruitful. Similarly, he who flies high and is raised farther from the earth will find more air beneath him to sustain him, and consequently will be less impeded by the weight of gravity; in fact, he will be able to fly so high that, having no difficulty in cutting through the air, he will not be able to redescend, even though one may judge it easier to cut through the air's depth toward the earth than the air above toward the stars.

CES.   So much so that with this sort of progress he acquires always more and more ease in raising himself?

MAR.   Exactly. And Tansillo also says:

> The more I feel the air beneath my feet,
> the more I spread proud pinions to the wind
> despise the world, and further my way to
> heaven.

The more every part of every body, including those of the elements, arrives nearer to its natural place, so much greater is its impetus and force, so that in the end willy nilly it must reach its destination. Thus, as we see that all the parts of bodies are drawn toward their proper places, so must we judge that things of the intellect are drawn toward their

---

[39] Here Bruno follows Marsilio Ficino who compares the soul's desire for God to fire moving upward, and its desire for earth to a stone falling downward. See FICINO, *Opera omnia* (Basel, 1561), p. 304.

proper objects as toward their own place, home and end. Now you can easily see the complete meaning intended by the emblem, the motto and the verses.

CES.   So much so that anything that you might add to it would seem to me most superfluous.

IX.

CES.   Let us see now what is represented by those two burning arrows upon a shield and by the above inscription, *Vicit instans*. [40]

MAR.   This emblem represents the war which continues in the soul of the frenzied one. Because of too long an intimacy with matter his soul was too stubborn and inert for penetration by the rays of the splendor of the divine intelligence and the species of the divine goodness; during all this time, he says, his heart was armored with diamond, meaning that its stubbornness and refusal to become heated and penetrated had protected it from the blows love brought him in its assaults from all sides. [41] He means that he has not been wounded by those blows of eternal life of which the Canticle speaks when it says, *Vulnerasti cor meum, o dilecta, vulnerasti cor meum*. [42] These blows are not caused by iron or other metal by means of some powerful and strenuous force, but are caused by the arrows of Diana or of Phoebus. Goddess of the wilderness where Truth is contemplated this Diana is the order of the secondary intelligences, who reflects the splendor of the first intelligence in order to communicate it to those who are deprived of its more direct vision. As for Phoebus, he is the principal god Apollo, [43] who with his own unborrowed splendor

---

[40] "The moment conquers."

[41] See above, pp. 164-165:

A better forger of Aetna, a better smith, anvil and hammer do I find here in this breast... where the soul lies prostrate from so many assaults of such long tortures and great martyrdoms...

[42] *Cant.* 4:9:

...Thou has wounded my heart, my sister, my spouse, thou has wounded my heart with one of thine eyes, and with one hair of thy neck...

[43] Apollo is universal intelligence and informing principle of the *anima mundi;* Diana is the universal material principle which receives the inform-

transmits his arrows, in every direction, that is, his rays, which are the innumerable species and marks of the divine goodness, intelligence, beauty and wisdom. The frenzies of love will depend upon the way these arrows are received; therefore the adamant subject may cease to reflect the light as it strikes him on the surface, and, on the contrary, softened and conquered by the heat and light, he may become entirely luminous in substance, may become himself all light, because his affection and his intellection have been penetrated. This does not happen at once at the beginning of life, when the soul freshly sets out inebriated of Lethe and is still full of the waters of forgetfulness and confusion; for there the soul is intimately a prisoner of the body and most concerned with the care of its vegetative life; but little by little the soul orders itself to become active in the exercise of its sensitive faculty, until that moment when by its rational and discursive powers it becomes more purely intellective. Then the soul can be raised to the mind and no longer feels beclouded by the murkiness of that humor which, thanks to the exercise of contemplation, is no longer putrefied in the stomach, but has been fully digested. [44]

In this disposition this frenzied one shows that he has endured six illuminations, in the course of which he has not yet arrived at the purity of concept which could have made him a fitting abode of those alien species which offer themselves equally everywhere and forever knock at the door of the intelligence. Finally love, who from many sides and on many occasions has assaulted him in vain (just as the sun is said to expend light and heat in vain for those who are in the bowels and obscure depths of the earth), *fixed itself in those sacred lights:* That is, love revealed itself under the two intelligible species of the divine beauty, bound his intellect by the light of truth, burned his affection by the light of goodness, and conquered the corporeal and vegetative ardors which

---

ing principle and communicates it to the lesser species. Apollo corresponds to the one eternal light of Plotinus, and Diana to Plotinus's *nous:* See PLOTINUS *Enneads* 4, 8, 6; 5, 7; also above, introduction, pp. 41-42.

[44] For a comparative account of the evolution of the soul's self-consciousness, see DANTE *Convivio* 3, 4; also *Purgatorio* 25, 67-107.

until that time had seemed to triumph and to remain intact (despite the excellence of the soul). For those lights which reflect the active intellect, the illuminator and intellectual sun, easily penetrated his own lights, the light of truth by the door of the intellectual potency, the light of goodness by the door of the appetitive potency down to his heart, that is, to the substance of the passion in general. This, then, was that *double arrow* which *came from the hand of the irate warrior* and was more prompt, efficacious and more ardent than it had been a little while ago when it had shown itself to be more feeble and neglectful. Thus, when that heat and light of truth illuminated his intellect for the first time, he experienced that victorious moment because of which it was said, *vicit instans*. Therefore you can understand the sense of the proposed emblem, motto and the sonnet which says:

Strongly I waxed in virtue under the blows
of love, when assaults from many and varied parts
were sustained by a heart armored with diamond.
Thus my efforts triumphed over those of love.

At last, one day (as the heavens destined
it) I found myself so fixed by those sacred
lights, which through my eyes, and alone among
all the others, found easy entrance to my heart.

Then was hurled upon me that double arrow,
which came from the hand of the irate warrior,
and for six illuminations had failed to assail
me. [45]

It pierced its mark, and there fixed itself
firmly, and planted its trophy upon me where it
could restrain my fugitive pinions.

And since then with more solemn preparation
the anger of my sweet enemy never ceases to
wound my heart.

---

[45] See BRUNO, *Opere Italiane*, ed. Gentile, II, 453, and note 2, for the statement that the six illuminations refer to the first thirty years of Bruno's life before 1578 when he was first inspired by a conversion to his new philosophy.

It was a single moment which marked both the beginning and the fulfillment of victory. It was a unique two-fold species, which alone among all other species found easy entrance; for in that two-fold species is contained the efficacy and virtues of all the other species; for what greater and more excellent form can be manifested than that beauty, goodness and truth which is the source of every other truth, goodness and beauty? The two-fold species *pierced its mark*, took possession of the heart, marked it, impressed its character there, *and fixed itself firmly;* then it established itself, confirmed itself and strengthened its position so that it could never be lost; for that reason it is impossible for one to turn to love anything else once he has received the divine beauty within himself; and it is impossible for him not to love it, as it is impossible that the appetite can reach out for anything other than the good or a species of the good. And this must be consummately in accord with the appetite for the highest good. Thus, *restrained are the pinions* which were formerly fugitive, accustomed to flying below with the weight of matter. Consequently, *the sweet anger never ceases to wound* the heart, soliciting the affection and reawakening the mind; for the *sweet anger* is the efficacious assault of the benignant enemy who had been excluded for such a long time as a stranger and an alien. And now that enemy is the sole and complete possessor and disposer of the soul; for the soul does not desire or wish to desire anything but him; nor is it content nor does it wish to be content with anything else, as the poet often says:

Sweet anger, delicious war, sweet darts,
Sweet are my afflictions and sweet are my pains. [46]

---

[46] PETRARCH *Rime* 205:

Dolce ire, dolci sdegni e dolci paci,
Dolce mal, dolce affanno, e dolce peso,
... ...
Sweet anger, sweet scorn and sweet tranquillity,
sweet evil, sweet anxiety, and sweet burden,
... ...

X.

CES.   It seems to me there remains nothing else to consider pertinent to that emblem. Now look at this quiver and bow. That these belong to love is demonstrated by the surrounding sparks, a suspended noose, and the motto, *Subito clam.*[47]

MAR.   I recall quite well having seen this expressed in the poem. But let us read it first:

Eager to find the prey he covets, the
eagle wings his way toward the sky, warning
all the animals that at his third flight he
prepares for destruction.

And from the deep cavern the vast roar
of the ferocious lion brings mortal terror,
so that the beasts, foreseeing the evil,
scurry their scant breakfast to their caves.

And when the whale leaves the caves of
Thetis to assail the mute herd of Proteus,
he first makes felt his violent spray.

The eagles of the ḻy, the lions of the
land and the whales who rule the sea do not
come treacherously; but the assaults of love
come in secret.[48]

Ah, for me those happy days were shattered
by the power of one instant, which made of me
an unfortunate lover forever.

There are three regions of animals and these are
composed of the major elements of earth, water and

---

[47] "Suddenly and secretly."
[48] PETRARCH *Rime* 2:

Per fare una leggiadra sua vendetta,
E punire in un dì ben mille offese,
Celatamente Amor l'arco riprese,
Come uom ch'a nocer luogo e tempo aspetta.

To make his revenge an artful thing,
and punish a million offenses in one day,
secretly love took up his bow, as one waits for
the time and place to strike.

air. These animals are of three genera; wild beasts of prey, fish and birds. Of these three genera nature has provided three chief species: the lion on land, the whale in the sea and the eagle in the air. Each one of these, as if to show that it has force and power superior to the other, will go so far as to behave with manifest magnanimity, or at least with a semblance of it. For that reason it is observed that before beginning the chase the lion sends out a powerful roar which makes all the woods resound, as the poet says of the frenzied hunter:

At saeva e speculis tempus dea nocta nocendi,
Ardua tecta petit, stabuli et de culmine summo
Pastorale canit signum, cornuque recurvo
Tartaream intendit vocem, qua protinus omne
Contremuit nemus, et silvae intonuere profundae. [49]

Whe know too that when the eagle wishes to seize its prey, it first flies from its nest toward the sky in a vertical and perpendicular position; but ordinarily, after the third time, it leaps up with great impetus and swiftness as if it would fly along a horizontal plane; in that manner, seeking the advantage of a swift flight and making use of the time to examine its prey from afar, it either rejects it or resolves upon it after having fixed its eye upon it three times.

CES. Can we conjecture the reason why it fails to attack its prey at once when it sees it for the first time?

MAR. Not precisely. But perhaps at this moment the eagle perceives it may be offered a better or an easier prey. Besides, I do not believe it always acts in this way, but only generally. Now to return to our discourse. With respect to the whale, we know that because it is a very large organism, it cannot cut through the waters without making its presence manifest beforehand by the reaction of the waves. Besides, there are found many other species of the same fish whose

---

[49] VIRGIL *Aeneid* vii 511-515:

...But the grim goddess, seizing from her watch-tower the moment of mischief, seeks the arduous roof, and sounds the pastoral signal from the highest summit of her abode, and strains her Tartarean voice on the twisted horn, which made the entire forest tremble, and echo through the deep wood...

movement and respiration exhale a windy and tempestuous spray of water. Therefore the inferior animals can take the time to escape from all three species of superior animals, so that these superior animals
do not behave as deceivers and traitors. But Love,
who is stronger and mightier than these animals, and
exercises supreme dominion in heaven, on earth and
in the sea, and, perhaps, like these animals, ought to
show a magnanimity the more excellent, the more
power it has, nevertheless directs its assaults unexpectedly and wounds suddenly:

Labitur totas furor in medullas,
Igne furtivo populante venas,
Nec habet latam data plaga frontem;
Sed vorat tectas penitus medullas,
Virginum ignoto ferit igne pectus. [50]

As you see, this tragic poet calls love *furtive fire,
unknown flame;* Solomon calls it *furtive water.* [51] Samuel named it a *murmuring of a subtle breath.* [52]
The three indicate the sweetness, suavity and cunning with which love comes to tyrannize over the universe on the sea, on land and in heaven.

CES. There is no larger kingdom, nor worse tyranny, no
better domain, no power more necessary, nothing
sweeter and more gentle, no food more sharp and bitter, no god more violent, none more amiable, no agent
more perfidious and more feigning, no author more
regal and faithful than love. And, finally, it seems to
me that love is everything and does everything, and
that everything can be said of it and everything can
be attributed to it.

MAR. You express it very well. Love, then (something
which acts principally through the vision, as through
the most spiritual of all the senses, for the vision as-

---

[50] SENECA *Phaedra* II. iii:

...Madness slides down into the innermost part of the veins by
a furtive, ravaging fire; and it does not wound the wide open
breast; but devours the disguised innermost marrow and destroys
the courage of virgins by an unknown flame...

[51] Prov. 9:17.
[52] III *Kings* 19:12.

cends immediately to the perceptible limits of the world and without delay extends itself to the farthest horizon of the visible) will be ready, furtive, unexpected and sudden. [53] Besides, we must consider that, according to the ancients, love comes before all the other gods; for that reason there is no need to invent a fable of Saturn who shows love the way, and then is forced to follow it himself. Moreover, why should it be necessary to see if love appears and announces itself externally, if its dwelling is in the soul itself and if its bed is the heart, and if it resides in the composition of our very substance, and is one with the impulse of our potencies? In conclusion, in all things the appetite for the beautiful and the good is natural, and for that reason it is unnecessary to argue or discourse to see how the affection is formed and strengthened; for suddenly and in a single instant the appetite is joined to the desirable, just as the vision is joined to the visible.

XI.

CES.   Now let us inquire into the meaning of that burning arrow about which the motto, *Cui nova plaga loco*, [54] is inscribed. What is this arrow's target? Explain this to me.

MAR.   That the burning terrors of Lybia and
Puglia destroy so much corn or commit so many
ears of wheat to the wind; that the orb of
the great star emits so many translucent rays;

that this soul, happy in its profound
pain and so sad in the joy of its sweet torment, receives burning darts shot from a
double star, all sense and reason forbid me
to believe.

---

[53] PLOTINUS *Enneads* 4. 3. 10; 2. 4. 5; 3. 6. 2.

[54] "Where does the new wound strike?" Bruno in the emblems of bow, arrow and noose follows Petrarch whose poetic imagery employs the beloved as an irate, sweet warrior who captures her lover unaware and strikes him with love's dart. See PETRARCH *Rime* 3, 21, 87. See also above, p. 204, for the motto, *vicit instans*, "the moment conquers".

What more do you attempt, sweet enemy,
Love? What zeal moves you to strike me with
new blows, now that my whole heart has become
one wound? [55]

Because neither you, nor any other force
has a single point left on which to strike
another blow, or a single point to pierce or
sting me, go, turn your bow elsewhere.

Cease wasting your effort here, for it
is wrong, if not vain, oh god of beauty, to
kill one who is already dead.

The entire sense of this poem is metaphorical as in
the case of the preceding ones, and it is in this sense
that it can be understood: the multitude of arrows
which wound and have wounded the heart, represent
the innumerable individual objects and species of
objects which, according to their degrees, reflect the
splendor of the divine beauty and therefore kindle
the passion for the desired and apprehended good.
Both the desired and the apprehended good, inasmuch
as the one is goodness in potency and the other is
goodness in act, and one is a possible and the other
an actual good, crucify and console at the same time,
and give at once a sense of the bitter as well as the
sweet. But when all the affections are completely
converted to God, that it, to the idea of ideas, by
the light of intelligible things, the mind is exalted
to the suprasensual unity, and is all love, all one, and
it no longer feels itself solicited and distracted by
diverse objects, but becomes one sole wound, in which
all the affections gather to become a single affection.
Then it is not the love or appetite of a particular thing
that can solicit or even approach the will; for there is
nothing more right than justice, nothing more beautiful
than beauty, nothing that has more goodness than the
good; nothing can be found greater than greatness
itself; nothing more luminous than the light which
by its presence obscures and effaces all other lights.

CES.   To the perfect, if it is perfect, there is nothing
that one can add; that is why the will is incapable of

---

[55] PETRARCH *Rime* 2, 133.

any other appetite when it experiences the supreme and sovereign perfection. I can therefore understand his conclusion, when he says to love, *cease wasting your efforts here; for, if not in vain,* it is wrong (according to a certain analogue and metaphor) *to try to kill one who is dead,* that is, one who is deprived of life and insensible to other objects, so that he can no longer be *stung* or *pierced* by them; for what would it profit him now to be exposed to any other species? And this lament befalls him who, having tasted of the ultimate unity, would become entirely delivered and cut off from the multitude.

MAR.   You understand it very well.

XII.

MAR.   Now here beside us is a boy in a boat who is at the point of becoming engulfed by the stormy sea and, faint and languishing, has abandoned the oars. The emblem bears the motto, *Fronti nulla fides.*[56] Undoubtedly this means that the serene aspect of the waters invited the boy to plough the faithless sea; whose surface became unexpectedly turbulent, and caused him extreme and mortal fear, and because of his inability to resist the impetus of the waves, he was forced to abandon himself, head down, arms stretched out, and all hope lost. But let us read the verse:

Gentle boy, who from the shore let loose
the tiny boat, and, longing for the sea, offer
an untutored hand to a frail oar, you are suddenly aware of your misfortune.

You see that the treachery of the baneful
sea, makes your prow sink too low or rise too
high; nor does your soul, overcome by importunate
desires, avail against the oblique and surging
billows.

Cede the oars to your fierce enemy, and
with less disquiet await your death; and that
you may not see death, close your eyes.

---

[56] "No faith in this face." See above, p. 197 and note 29. See also JUVENAL *Satires* II, 8.

If some friendly aid is not prompt, any
moment you will surely feel the ultimate effect
of your most ignorant and curious zeal.

My harsh destinies are comparable to
yours, because, longing for Love, I experience the rigor of that lord of traitors.

How and why love is a traitor and fraudulent we
have seen a little while ago. But because I see that
the following poem is without an emblem and motto,
I suppose it might be related to the preceding one.
Therefore let us read it:

Having left the shore to try myself and
relax a little while from my sober labors, I
fell to musing almost playfully, when suddenly
I saw the cruel fates.

These have burned me with so violent a
fire that in vain do I attempt the more secure
shores again, and in vain do I invoke for deliverance a hand of mercy which would promptly
carry me aloft to my swift enemy.

Impotent to release myself, hoarse and
vanquished, I yield to my destiny, and no longer
try to build a useless bulwark against death.

May my cruel destiny deliver me from every
other life, and prolong no more the final torment
which it has prescribed for me.

Exemplar of my great evil is the improvident
boy who abandoned himself as a plaything to
the bosom of the enemy.

At this point I am not certain that I understand or
explain everything the frenzied one means. However
one thing that is very clear is the strange condition
of a soul discouraged on the one hand by the awareness
of the difficulty of the work, by the great amount of
fatigue and the vastness of the undertaking, and on
the other hand discouraged by its own ignorance, its
lack of skill, weakness of nerves and the danger of
death. He is without counsel for his undertaking; he
does not know where he must turn or to whom; he

perceives no place of flight or of refuge, for the waves menace him from all sides with their frightening and mortal assaults. *Ignoranti portum nullus suus ventus est.* [57] This lover realizes he has relied too much on his own good fortune, having prepared for himself only turmoil, captivity, ruin, submersion. He sees how fortune sports with us; the gifts with which she gently fills our hands she causes to fall and break, or she sees that they are taken from us by another's violence, or she makes them suffocate, poison, or disquiet us by arousing in us suspicion, fear and jealousy to our great loss and ruin. *Fortunae an ulla putatis dona carere dolis?* [58] Because strength that cannot prove itself is vain, magnanimity of soul that cannot prevail is nothing, and because labor that bears no fruit is useless, he sees the effect of the fear of evil, which is worse than the evil itself. *Peior est morte timor ipse mortis.* [59] Because of fear he already suffers everything he is afraid to suffer: trembling of the limbs, weakness of the nerves, tremors of the body, anguish of the spirit; and he brings upon himself what has not yet befallen him, a thing certainly worse than whatever could overtake him. For what is more witless than to bemoan something in the future, which is not felt in the present?

CES.   These considerations explain the superficial aspect and external iconography of the emblem. But it seems to me the argument of the frenzied one refers to the weakness of the human mind, which, completely engaged in the divine enterprises risks finding itself suddenly engulfed in the abyss of an incomprehensible excellence; and therefore the sense and imagination become confused and absorbed, so that not knowing where to turn, equally incapable of going forward or turning back, the human mind vanishes and loses its own existence like a drop of water that loses itself in the sea, or a weak breath dissipated as it loses its substance in the spacious and immense atmosphere.

MAR.   Good, but let us go now, and discuss it on the way home, for it is getting dark.

END OF THE FIRST DIALOGUE

---

[57] "To one ignorant of the port, there is no wind to guide him."
[58] "Do you think any gift of fortune is without pain?"
[59] "The fear of death is worse than death itself."

XIII.

MAR.

Here is a flaming yoke enfolded by a noose, and around it the inscription, *Levius aura*.[1] The emblem means that divine love does not oppress or lead its servant to the shades below as a captive and a slave, but raises, uplifts and exalts him beyond every freedom.

CES.

I beg you, let us read the poem quickly; then in better order, more precisely and with no delay shall we be able to examine its sense and see if we can find even another meaning in it.

MAR.

It says:

> She who kindled my mind to the higher love, she who rendered every other goddess base and vain to me; she in whom beauty and sovereign goodness are uniquely displayed,

> is she whom I saw coming from the forest, huntress of me, my Diana, among the lovely nymphs upon the golden Campania, wherefore I said to Love: —I surrender myself to this one.

> And he to me: —Oh fortunate lover! Oh spouse favored by your destiny! She who alone among so many

> has within her bosom life and death, and adorns the world with holy graces, her you have achieved by labor and by fortune;

---

[1] "Lighter than the air." See above, introduction, p. 53 and notes 16 and 17, and p. 55.

captive though I am in her amorous court, I am so highly blessed, that I do not envy the freedom of any man or god.

You notice how content he is under such a yoke, under such a burden, captive of the one he saw proceed from the forest, from the wilderness and from the wood; that is to say, from those less frequented regions ignored by the multitude, alien to society and apart from the vulgar. Diana, splendor of the intelligible species, is his huntress, because having wounded him by her beauty and grace, she has bound him and holds him under her sway more content than he could have ever been otherwise. She is said to be *among the lovely nymphs,* that is to say, among the multitude of other species, forms and ideas, and *upon the golden Campania,* an allusion to that intelligence and spirit that appears in Nola, and lies on the plain of the Campanian horizon. To her he renders himself, to her whom love praised more than he praised any other, desiring that he regard himself most fortunate because of her, who, among all that is visible and invisible to the eyes of mortals, gives the world its noblest attire and makes man glorious and beautiful. That is why he says his mind is enkindled to that highest love and that it recognizes *every other goddess,* that is, the care and consideration of every other species, as base and vain.

Now in proclaiming that his mind has been kindled by the highest love, he offers us an example of how to raise the heart as high as possible by our thoughts, labors and works, and how not to divert ourselves with things base and inferior to our faculty, as happens to those who either because of avarice, negligence, or even from some other unfitness, remain in this brief span of life attached to ignoble things.

CES. It is necessary that there be artisans, mechanics, farmers, servants, pedestrians, the ignoble, the base, the poor, the pedants and others of the sort; for otherwise there could not be the philosophers, saints, educators, lords, captains, noblemen, illustrious men, wealthy men, wise men and others who are as heroic as are the gods. Why then, ought we to be forced to corrupt the law of nature which has divided the universe into things that are greater, and things that are less, things superior and things inferior, things illuminating and things obscure, things worthy and unworthy, not only outside of us, but also within us, in our very own substance, even to that part of our substance affirmed as immaterial? It is the same among the intelligences; some are inferior and others are superior, some serve and obey, while others command and govern. But I do not hold that this ought to serve as an example by which the order of things should become perverted and confounded because subjects wish to become rulers and the ignoble wish to become noble with the result that finally a certain state of neutrality and bestial equality would follow, a condition one finds in certain solitary and uncultivated republics. Besides see what damage has come to the sciences because the pedants have wished to become philosophers, and while treating of the things of nature have meddled in determining things divine? Who does not realize that harm has come and still comes because not all minds are equally kindled to the highest love? Who has good sense and does not see the profit reaped by Aristotle, Alexander's master of letters, when he used his noble intellect to contradict and make war upon the Pythagorean theory and the theory of the natural philosophers? By the process of logical reasoning he wished to offer definitions, notions, certain quintessences and other fragments and miscarriages of fantastic thought as though they were the principles and the substances of things, more concerned as he was with the opinions of the mob and the stupid multitudes who are guided and lead more by means of sophisms and the superficial appearances of things than by the truth hidden in the substance of them, a truth which is the very substance of those things. He alerted his mind not to contemplate but to judge and give an opinion about things he had never studied and of things of which he had not even heard. Therefore so much of the good and of the rare which he offers from the matter of his poetics, logic and metaphysics, in our day in the hands of other pedants who labor with the same *sursum corda* becomes formulated in new dialectics and modes of forming the reason, modes inferior to the doctrine of Aristotle, just as perhaps the doctrine of Aristotle is imcomparably inferior to that of the ancients. This has already happened because certain grammarians, having worn themselves out upon the rumps of infants and on the anatomies of words and phrases, have wished to set their minds to the creation of a new logic and meta-

physics, judging and giving opinions about matters
they have not hitherto studied and do not understand
now. That is why by the favor of the ignorant multitude (to whose wit they more conform) these grammarians can so well give the final blow to the letters
and observations of Aristotle, just as Aristotle himself
was the hangman of other divine philosophers. See
then what ordinarily results from the advice that
everyone should pretend to aspire to the holy light
and hold all other emprises base and vain.

MAR.

> Ride, si sapis, o puella, ride,
> Pelignus, puto, dixerat poeta;
> Sed non dixerat omnibus puellis;
> Et si dixerit omnibus puellis,
> Non dixit tibi. To puella non es. [2]

Therefore the *sursum corda* is not meant for everyone, but only for those who have wings. We see quite
well that pedantry has never been more exalted for
governing the world, than in our day; [3] and it opens
toward the true intelligible species and objects of the
one infallible truth as many paths as there are pedants.
For that reason in this age well born intellects must
be awakened to the greatest extent, armed with the
truth and illumined by the divine intelligence, in order
to take up arms against the darkness of ignorance and
to ascend that high rock and eminent tower of contemplation. These are the intellects which must hold
every other enterprise as vile and vain.

These intellects must not waste time, whose speed
is infinite, on things superfluous and vain; for with
astonishing speed the present slips by and the future
approaches with equal rapidity. What we have endured
is nothing, what we endure now is a point, and what

---

[2] MARTIAL *Epigmms* II, 1, 1-5:

> Smile, if you are wise, maiden, smile,
> Paelignus, the poet said, I believe;
> But he spoke not to all the maidens;
> And indeed had he spoken to all the maidens,
> He did not speak to you. For a maiden you are not.

[3] For the author's personal battle with "pedantry" see DOROTHEA SING-
ER, *Bruno, His life and Thought*, pp. 32-33.

we shall have to endure is not even a point, but can become a point which at the same time will be and will have been. And still one man encumbers his memory with geneology, another attends to deciphering ancient writings, and still another is occupied with multiplying the sophisms of children. You will see, for example, volumes filled with reasoning such as:

> Cor est fons vitae,
> Nix est alba;
> Ergo cornix est fons alba.

One warbles about whether the noun existed before the verb; the other about whether the sea existed before its source; another desires to revive obsolete words—because an ancient writer once employed them he would raise them again to the clouds; another obsesses himself with false and true orthography; and still others preoccupy themselves with similar nonsense, more worthily scorned then heeded. For this they fast, become lean, grow consumptive, let their skin dry up, their beards grow, putrefy, and upon this throw down the anchor of the highest good. In the name of these futilities they scorn fortune and by them they build a rampart and a shield against the thrusts of fate. By the grace of these vile notions they think they ascend to the stars and are like the gods, and they think they comprehend the beautiful and the good which philosophy promises.

CES.   It is amazing indeed that time, which can not suffice us for things that are necessary, no matter how diligently we guard it, becomes more often wasted on superfluous things, in fact upon things vile and shameful.

It is no laughing matter that the following is attributed to Archimedes (or to certain others who follow him) as a laudable action. At the moment when the city was in ruins and people were scurrying in all directions, when his room was on fire, his enemies in his chamber and at his back, at whose discretion and whim lay the loss of his skill, brain and life, despite all this, he nevertheless lost the instinct and desire for self preservation and forgot everything in order to find the proportion between the curve and the straight line, the diameter and the circumference of a circle or to solve some other similar problem, all worthy of youths, but unworthy of one who, if he could, should have grown old intent upon things more worthy of the goal of human study.

MAR.   I approve of what you yourself said a little while ago about this subject, that the world must be full of all sorts of people and the number of imperfect, ugly, poor, unworthy and nefarious ones must be in the majority; in conclusion, it ought not be otherwise than it is. The long life of Archimedes, Euclid, of Priscian, of Donatus and of others, who until their deaths were occupied with numbers, lines, verbal forms, grammatical convention, orthography, dialectics, syllogisms, methods, modes of thought, rudiments of speech and other isagoges, has been ordained for the profit of youth and children, who may learn and receive the fruits of the mature years of those men; fruits which they may eat appropriately in their green age, so that once adult they may find themselves apt and prepared for greater pursuits without difficulty.

CES.   I still maintain what I said a little while ago about those who on the one hand, labor to purloin the position and reputation of the ancients by producing new works, inferior or no better than those already produced, and spend their lives observing the skin of a goat or the shadow of an ass, and others who, on the other hand, as long as they live, labor to excel in exercises fit for children, and these for the most part without profit to themselves or to anyone else.

MAR.   Now we have said enough about those who either cannot or ought not presume to have *the mind kindled to the higher love.* Let us consider now the voluntary captivity and delightful yoke beneath the sway of the mentioned Diana; I mean that yoke without which the soul is incapable of ascending to the loftiness from which it fell; for that yoke renders the soul lighter and more agile, and the noose gives it greater dispatch and liberty.

CES.   Then explain.

MAR.   To begin, continue and conclude in order, I consider that everything that lives, in whatever mode it lives, must in some manner nourish and feed itself. But to the intellectual nature only intellectual food is necessary, just as to the body only corporeal food is necessary; for nourishment is taken for no other purpose than to be absorbed into the substance of the thing nourished. Besides, the body can no more be transmuted into spirit than the spirit into the body; for a transmutation is possible only if the matter previously in the form of the one passes over to the form of the other; but the spirit and the body do not have a common matter which makes it possible for the subject of one domain to become the subject of the other.

CES.   Surely if the soul drew nourishment from the body, it would bear itself better where it found an abundance of matter (as Iamblicus argues), [4] so that when we see a big and fat body, we may believe it to be the vehicle of a valiant soul, firm, ready, heroic, and say, oh fat soul, oh fecund spirit, oh beautiful mind, oh divine intelligence, oh illustrious intellect, oh blessed hypostasis which would make a banquet for lions, or for dogs. In the same way an old man appearing half-decayed, weak and diminished in strength, would have to be deemed of little spirit, discourse and reason. But continue.

MAR.   The nourishment of the spirit, then, can be only the thing the spirit has always longed for, searched for, embraced and relished more willingly than any thing else, an object through which the soul is fulfilled, pleased, benefited and grows; and that object is the truth toward which man aspires at every moment, in every age, and in whatever condition he finds himself, and for which he usually scorns all fatigue, undertakes every zeal, counts his body for nothing and holds this life in contempt. For the truth is something incorporeal; and no truth, whether it be physical, metaphysical, or mathematical is found in the body, for you know very well that the eternal human essence is not to be found in the individuals who are born and die. It is the specifically one, Plato said, not the numerical multitude, which bears the substance of things. For that reason he calls the idea one and many, stable and mobile; because as incorruptible species it is intelligible and one; and as it communicates itself to the corporeal and is subject to motion and generation, it is something sensible and many. In this second mode it has more of non-being than of being, for it is always one thing and another and its privation imposes an eternal course

---

[4] Pseudo-Jamblicus *De mysteriis*, in MARSILIO FICINO, *Opere*, p. 1876.

upon it. You see, moreover, that the mathematicians
have agreed that perfect figures are not found in
natural bodies, and they cannot exist either by the
power of nature or art. Besides, you know that the
truth of supersensual substances is beyond the corporeal.

One concludes, then, that he who seeks the truth
must ascend above the order of corporeal things. Besides, it must be considered that everyone who is
nourished has a certain notion and natural memory of
his food, and always (especially when his nourishment
becomes more necessary) retains the similitude and
species of that food, and retains it the more nobly,
the more noble he is who seeks, and the more glorious
the object sought. Every one has innate knowledge of
things which assure the conservation of his individu-
ality and his species, and therefore his ultimate perfection; and this is the reason why every being in-
dustriously seeks nourishment through some species of
prey.

Thus it is necessary that the human soul have the
light, the ingenuity and instruments adopted to possess
its own prey. Toward such an end the contemplation
gives assistance and toward this end logic is used,
the organ most adept for the acquisition of the truth,
for distinguishing, exploring and making judgments.
Then the soul will proceed to traverse the forest of
natural phenomena where so many objects are hidden
under a shadow and cloak; for in a thick, dense and
deserted solitude the truth voluntarily seeks cavernous
retreats, interwoven with thickets and surrounded by
wooden, rugged and leafy plants, and there for the
most worthy and excellent reasons she conceals, veils
and buries herself with the greatest vigilance; just
as we are accustomed to conceal most diligently our
greater treasures, so that the multitude and variety
of hunters (some having more skill and practice than
others) cannot discover them without great pain. To
that forest Pythagoras proceeded, seeking the truth
by following its traces and vestiges in nature, that
is, in the numbers which in a certain way make the
progress, considerations, modes and operations of the
truth apparent; for it is in number insofar as it applies
to the many, to measurements, to time and to weight
that the truth and essence of all things is found. There
Anaxagoras and Empedocles proceeded, who, considering that the omnipotent and omnipresent divinity en-

compassed the universe, found nothing so minute which could not have the divinity concealed beneath it, in accordance with every argument; yet they never failed to proceed to that region in which the divinity was predominant and expressed by the most noble and magnificent argument. There the Chaldeans searched for the divinity by way of abstraction, not knowing what to affirm about it; and they advanced without demonstrations and syllogisms, and tried to penetrate further by brushing aside obstacles, furrowing the field and clearing the forest, by a forceful denial of every species and predicate whether comprehensible or secret. Plato searched for it by alternately tearing down and building up barriers, so that the inconsistent and fleeting species would remain as in a network held in a row of definitions; for he considered that superior things exist by participation, similitude and reflection in inferior things, and that inferior things according to their greater degree of dignity and excellence exist by their participation in superior things; and he considered that the truth is in the one and the other according to a certain analogy, order and scale in which the lowest degree of the superior order joins the highest degree in the inferior order. In this way, by traversing the intermediary degrees, he contributed a progression from the lowest in nature to the highest, a progression from evil to good, from darkness to light, from pure potency to pure act. Even Aristotle boasted of being able to arrive at the desired prey by means of the footprints and vestiges that could be traced when from effect he wished to reascend to cause. However most of the time (and more than all the others who preoccupied themselves in such a chase) he lost the way, hardly knowing how to distinguish between the vestiges.

Finally, some theologians, nurtured in the doctrines of various sects, seek the truth of nature in all its natural and specific forms; and they consider that it is through these forms that the eternal essence specifically and substantially perpetuates the everlasting generation and mutation of things called into existence by those who create and build them; and that over those who build them reigns the form of forms, the source of light, the truth of truths, the god of gods, by whom everything is filled with divinity, truth, being and goodness. Therefore truth is sought as something inaccessible, an object beyond objectivity and beyond all comprehension. For that reason it is impossible for anyone to see the sun, the universal Apollo and absolute light as the supreme and most excellent species; but very possible to see its shadow, its Diana, the world, the universe, the nature which is in things, the light shining through the obscurity of matter and so resplendent in the darkness. Therefore of all those who in the ways mentioned speculate much in this deserted wood, very few are those who arrive at the font of Diana. Many remain happy with chasing the wild and less illustrious beasts, and most of them find nothing to catch, for they have aimed their nets at the wind, and have remained with a handful of flies. I say very few are the Actaeons to whom destiny gives the power to contemplate Diana naked, and the power to become so enamoured of the beautiful harmony of the body of nature, so fallen beneath the gaze of those two lights of the dual splendor of goodness and beauty, that they are transformed into deer, inasmuch as they are no longer the hunters but the hunted. For the ultimate and last end of this chase is the capture of a fugitive and wild prey, through which the hunter becomes the hunted, the pillager becomes the pillaged. Because in all the other species of the chase undertaken for particular things, it is the hunter who seeks to capture those things for himself, absorbing them through the mouth of his particular intelligence; but in that divine and universal chase he comes to apprehend that it is himself who necessarily remains captured, absorbed, and united. Therefore, from the vulgar, ordinary, civil and ordinary man he was, he becomes as free as a deer, and an inhabitant of the wilderness; he lives like a god under the protection of the woods in the unpretentious rooms of the cavernous mountains, where he contemplates the sources of the great rivers, vigorous as a plant, intact and pure, free of ordinary lusts, and converses most freely with the divinity, to which so many men have aspired, who in their desire to taste the celestial life on earth have cried with one voice, *Ecce elongavi fugiens, et mansi in solitudine.* [5]

---

[5] Pss. 54 : 8:

Lo, I have gone far off flying away; and I abode in the wilderness.

The result is that the dogs, as thoughts bent upon divine things, devour this Actaeon and make him dead to the vulgar, to the multitude, free him from the snares of the perturbing senses and the fleshly prison of matter, so that he no longer sees his Diana as through a glass or a window, but having thrown down the earthly walls, he sees a complete view of the whole horizon. And now he sees everything as one, not any longer through distinctions and numbers, according to the diversity of the senses, or as varied fissures are seen and apprehended in confusion. He sees the Amphitrite, the source of all numbers,[6] of all species, the monad, the true essence of the being of all things; and if he does not see it in its own essence and absolute light, he sees it in its germination which is similar to it and is its image: for from the monad, the divinity, proceeds this monad, nature, the universe, the world; where it is contemplated and gazed upon as the sun is through the moon, which is illuminated by it, inasmuch as he finds himself in the hemisphere of intellectual substances. She is Diana, she who is the being and truth of intelligible nature, in which is infused the sun and the splendor of a superior nature, according as the unity is distinct in that which is generated and that which generates, or that which produces and that which is produced. Therefore you will be able to draw your own conclusions about the mode of the chase, the dignity of the hunter and the most worthy result of his effort. That is why the frenzied lover boasts of becoming the prey of Diana to whom he renders himself, of whom he is esteemed a worthy consort, and so happy a captive under his yoke, that he has no reason to envy any man. For no other man has been given so much advantage as he. Nor has he reason to envy any god. For the species of a divinity cannot be obtained by an inferior nature, and consequently must not be desired, or even become the object of our appetite.

---

[6] Bruno in the Latin work, *De minimo*, I, defines the Amphitrite as "God, the Monad and source of all numbers". However, in the ensuing passages the Amphitrite is not God in his own proper essence, but God reflected in the universe and in nature. See JORDANI BRUNI NOLANI, *De Minimo*, in *Opera latine conscripta*, F. Fiorentino (Neapoli, 1879-1891), I.

CES.   I have understood well what you have said, and have been more than moderately satisfied. Now it is time to return home.

MAR.   Agreed.

END OF THE SECOND DIALOGUE

THIRD DIALOGUE

INTERLOCUTORS

LIBERIO   LAODONIO [1]

LIB.   While the frenzied one lay beneath the shadow of a cypress tree, and other thoughts allowed his soul to relax somewhat (a remarkable thing), it happened that his heart and his eyes (as though they were living beings and separate substances whose sense and reason were distinct from each other) engaged in a debate; [2] and each one complained that the other was the cause of the laborious torment that consumed his soul.

LAO.   If you remember their arguments, tell them to me.

LIB.   The dialogue was begun by the heart, which let the following accents burst forth from the depth of its breast:

FIRST ARGUMENT OF THE HEART TO THE EYES [3]

How is it, eyes of mine, that I am tormented so powerfully by that ardent flame which derives from [you?

---

[1] Little or nothing is known of Laodonio, and of Liberio there is also only scant information. However, Gentile and Spampanato suggest that Liberio is an allusion to a hamlet near Nola called Liveri, inasmuch as Nolans often were designated according to the geographical regions from which they came. See BRUNO, *Opere*, Gentile, II, 475 and note 1, and VINCENZO SPAMPANATO, *Vita*, p. 51 and note 4.

[2] See "Argument of the Nolan", pp. 70, 71 and note 16.

[3] The heart belongs to the material principle of the soul; the eyes to the informing principle. See BRUNO, *De la causa*, in *Opere*, I, pp. 194-

How can my mortal substance continue to be fed
by so great a fire,

that I believe all of the ocean's moisture and the
most frozen part of the slowest star of the Arctic to be
inadequate to curb my fire even for a moment and
[give me a
shadow of refuge?

You made me captive of a hand that holds
me, yet wants me not; because of you I am at
once buried in the body and exposed to the sun.

I am a principle of life, and yet, there is no
life in me. I do not know what I am, for I belong to
this soul, yet it does not belong to me.

LAO.   Understanding, knowledge and vision enkindle the
desire, and therefore through the ministry of the eyes
the heart becomes inflamed. The more lofty and
worthy the object that presents itself to the eyes, the
more powerful the fire and the more blazing the flames.
Now what object could so enflame the heart that it
dares not hope the coldest and most distant star of
the arctic can temper its ardor, nor hope that all the
waters of the ocean can appease its flames? How
excellent must the object be to have made the heart
an enemy of its own self, a rebel against the soul, and
contented in such enmity and rebellion, the captive of
a hand that scorns it and wants it not? But tell me
whether or not the eyes reply and what they have
to say.

LIB.   The eyes, on the other hand, complain against the
heart for having been the principle and cause of the
tears they have shed.
They reply to its lament with the following complaint.

FIRST REPLY OF THE EYES TO THE HEART

How is it, oh heart, that you pour forth
waters as great as the sea from which the

---

204 and II, 372 for the distinction, yet inseparability of the formal and
material principles in the infinite universe and in the individual soul.

Nereids ever raise their heads who die and are
reborn every day in the sun? Like Amphitrite,
the two-fold font,

(you) [4] can pour forth such immense rivers upon
the world, that you may say the river overflowing Egypt becomes a meager stream flowing into
the sea through seven double shores.

Nature provided twin lights to govern
this tiny world. But you, perverter of that
eternal order,

turned them into everlasting rivers.
And the heavens allow nature to be violated
and violence to endure.

LAO.   Naturally, fire and affliction in the heart cause
the eyes to fill with tears; and, of course, if the eyes
enkindle the flame in the heart, it is the heart that
causes the eye to fill with tears. [5] But I marvel at so
great an exaggeration, when the eyes say that the
heads of the Nereids do not emerge to the sun bathed
in more abundant waters. And besides these waters
are compared to the ocean not because they are diffuse, but because their two sources are able to pour
forth so many kinds of rivers, that compared to them
the Nile would appear as a small inlet divided into
seven streams.

LIB.   Do not be surprised at this exaggeration or at this
potency deprived of its act, for you shall understand
it all when you have heard the conclusion of this
argument. Now hear how the heart first replies to the
complaint of the eyes.

LAO.   I beg you, let me hear it.

LIB.   THE HEART'S FIRST REPLY TO THE EYES

Eyes, if an immortal flame is ignited in
me, and I am nothing else but a blazing fire;
if everything that approaches me burns up in

---

[4] Parenthesis mine.
[5] CATHARINE OF SIENA *Dialogues* 88.

smoke, so that I even see heaven burning in
my flames,

why does my great fire not consume you,
but produce in you a contrary effect? Why do
I moisten you and not burn you instead, if
fire and not moisture is my substance?

Blind ones, do you believe a two-fold
stream derives from so ardent a fire and that
those two living streams

derive their elements from Vulcan—as
sometimes of two contraries the one acquires
force, if the other resists?

See how impossible it is for the heart to persuade
itself that from one contrary cause and principle proceeds the force of a contrary effect; it goes so far as
to refuse to admit any such possibility, even by way
of *antiperistasi*. [6] This word refers to the vigor acquired
by one contrary while it flees the other contrary and
becomes united, self-enveloped, condensed and con-
centrated toward the individual substance of its own
virtue, which gains in efficacy what it loses in extension.

LAO.   Tell me how the eyes reply to the heart.

LIB.   THE EYES' FIRST REPLY TO THE HEART

Oh heart, your passion so confounds you, you have
lost the way to all truth. Whatever is revealed
or concealed in us has its origin in the seas.
Therefore, from us

and from nowhere else Neptune must be able to
recover his vast empire should fate decree to take it
from him. How can we be the source of your ardent
flame, we who are the twin parents of the sea?

Are you so mad as to believe that fire traverses
us, leaving behind it these two watery portals,

---

[6] See ARISTOTLE *Physics* 215a 15, 267a 16 for "the return of a counter
blow".

so that you might feel its immense flame?
Will you believe, as light penetrates glass, that fire
penetrates us?

It is not my intention here to philosophize upon the
coincidence of contraries, which I have worked out
in my book, *Of the Principle and the One.*[7] I will
suppose what is commonly supposed, that the contraries in the same category are as far apart as possible;
thus we shall more easily understand the sense of this
reply in which the eyes call themselves the origins or
fonts in whose virtual potency is the sea; so that, from
their potency, should Neptune lose all the waters of
the ocean, he could recall them into action, for they
are in that potency as in their principle and material
agent. However, when the eyes say that the flame
cannot pass through their rooms and portals to the
heart leaving so much water behind it, their argument
is not without reply, and this is true for two reasons.
First, because such an impediment could not actually
be present unless certain barriers were set up which
were actually insurmountable; second, because if the
waters were actually in the eyes, they could make
way for heat just as they could for light. For experience
shows that without burning the mirror a reflected ray
will light a material object exposed to it; moreover,
a ray of light will pass through a pane of glass, a
crystal, or a vase full of water, illumine the thing
it strikes and will not burn the liquid mass it has
traversed; thus is it a similitude and even true that
light produces impressions of dryness and burning in
the concavities of the deep sea. Consequently, by a
certain similitude, if not by an analogous consideration, one may see how it is possible that through the
deceptive and obscure organ of the eyes the affection

---

[7] Bruno in *De la causa, principio et uno* considers the material and
the formal equally substantial principles in the infinite universe and in the
individual soul. Accordingly, he conceives them to be contrary forces which
demand reconciliation for the attainment of total harmony. See BRUNO, in
Opere, I, pp. 186-189, 190-204. Bruno follows Nicolaus Cusanus who holds
the *coincidentia oppositorum* is fulfilled totally only in the divine being.
However Bruno finds the unity-of-opposites fulfilled considerably also in the
world of phenomena. See NICOLAUS CUSANUS, *Vision of God,* trans. G. E.
Salter (London, 1928), pp. 60-62, and *De docta ignorantia,* in *Opera omnia,*
ed. Hoffman and Klibansky (Lipsiae, 1932), i, 4, 12, 13; i, 26.

will be enkindled and enflamed by a light which does
not produce the same effect wherever it penetrates.
For the action of the sun's light as it traverses the air
is one thing, another as it approaches the senses, another as it penetrates everyone's sense, and still another
as it penetrates the intellect; and thus it proceeds
from one mode to another mode of being.

LAO.   Does the debate between the heart and the eyes
continue?

LIB.   Yes, because the eyes and the heart try to discover how it is that the heart contains so many flames
and the eyes so much water. Therefore, the heart
makes its second demand.

THE HEART'S SECOND ARGUMENT

If all the rivers run their course toward
the foamy sea and proceed to fill the blind
abyss, how is it, oh my eyes, that a two-fold
torrent proceeding from you is not discharged
upon the world

to extend the reign of the sea gods, dimin-
ishing the glorious charge of the other deities?
Why may one not see again the day when Deucalion
returned to his mountains? [8]

Where are the many overflowing rivers?
Where is the torrent to extinguish my flame, or,
if not to extinguish it, to enrage it the more?

Does not one drop descend to earth to diffuse itself there, that I may be allowed to
doubt what my appearance obliges me to believe?

What kind of potency is this that does not translate itself into act? This is what it would know. If the
waters are so numerous, why does not Neptune come
to tyrannize over the power of the other elements?
Where are the overflowing rivers? Where is the fresh-
ness fitted to cool the ardor of my flame? Is there not
one drop from the eyes to permit me to affirm what all

---

[8] OVID *Metamorphoses* I, 312-345.

appearance denies? But the eyes, in their turn, have
another question to ask.

THE SECOND ARGUMENT OF THE EYES TO THE HEART

If all matter is converted to fire and
then, like fire, mobile and light, is raised
to the lofty heaven, how is it that

tormented by so great a fire of love you
are not swept away swiftly as the wind in one
instant to the sun? Why do you wander a pilgrim
here below, and not find the path toward us
through the air?

No spark is seen flashing forth from that
breast; nothing appears which resembles a body
singed or reduced to ashes,

no smoke rises upward to make us weep:
each faultlessly guards its own state; and
neither the reason, sensation nor thought
are enflamed.

LAO.   This argument has the same value as the one before it, no more, no less. But let us come now to the
replies, if there are any.

LIB.   There certainly are and they are full of substance.
Listen.

SECOND REPLY OF THE HEART TO THE EYES

He is foolish who believes only in
appearances, and will not believe his reason;
my fire cannot take flight and no infinite
flame is seen, because

the ocean of the eyes has descended upon
it, and one infinite does not exceed the other. [9]

---

[9] i. e., the two opposing forces; form and matter, symbolized by water
and fire, eyes and heart, while distinct, are equally infinite and accordingly
demand reconciliation. See above, p. 228 and note 3.

If the fire and the sphere are counterbalanced,
it is because nature does not wish all to perish.

Tell me, by heaven, oh my eyes, which
path shall we ever take thanks to which you
or I will be able to render apparent the
cruel fate of our soul, that it may be rescued?

If our torments remain concealed, how
shall we render this god of beauty merciful
to us?

LAO.　　If this argument is not true, it is most original;
and if not original, it is excused in any case; for when
two forces exist, one of which is not stronger than the
other, both forces must stop functioning; because
the resistance of one is equal to the persistence of
the other, inasmuch as the one can attack as much as
the other can repulse the attack. Therefore, if in the
eyes the ocean of tears is infinite and the force of tears
is infinite, they must forever manifest themselves by
setting aflame or fanning the impulse of the fire hidden within the breast, and the eyes will never be
able to dispatch their twin currents to the sea, if the
heart puts an obstacle of equal force in their way.
This is why no appearance of tears flowing from the
eyes or flames flashing forth from the heart can invite
the beautiful deity to show mercy to the afflicted
soul. [10]

LIB.　　Now observe the following reply of the eyes:

SECOND REPLY OF THE EYES TO THE HEART

Ah, the impetuous force of our fonts is
wholly vain to pour forth their rivers to the
sea, for a contrary power keeps them hidden, so
that they send no rolling waters below.

---

[10] Only when the individual reconciles eventually the opposites of matter and form in his soul will the deity favor him. Then the fullness of being
he has achieved will make it possible for him to contemplate Diana, "the
highest monad", and unity-in-plurality. See above, introduction, pp. 42-43.

The infinite vigor of the burning heart
denies passage to the torrents that are only
too high; thus, our two-fold stream does not
flow into the sea, for nature abhors an earth
submerged.

Tell me, now, afflicted heart, you who
can oppose us with another force as great,
who would ever boast

of being the herald of so hapless a love
as ours, if your woe and ours can be so much
the less useful, the greater it is?

Just as two contraries of equal force are neutral-
ized, one and the other evil, being infinite, cancel
out; and such could not be the case were both of the
contraries finite, for in the natural order a perfect
parity is never realized, nor would such be the case
if one contrary were finite and the other infinite, for
the infinite contrary would certainly absorb the one
which was finite, and both contraries would manifest
themselves, or at least one would manifest itself by
the other. I leave the natural and moral philosophy
concealed beneath these statements to be sought, considered and understood by him who will and can. But
one thing I will not omit, that not without reason is
the heart's passion called an infinite sea by the apprehension of the eyes. Because the object of the mind
is infinite and no definite object is proposed to the
intellect, the will cannot be appeased by a limited
good. Beyond this good the will finds a still higher
good for itself, which it then desires and seeks, for,
as it is commonly said, the highest of the inferior species is also the lowest and first of the superior species,
whether this gradation ascends according to forms
(whose infinity we cannot estimate), or according to
the modes and reasons of those forms; and the highest
good being infinite, we believe it communicates itself
infinitely according to the condition of the things in
which it is diffused. Therefore, no definite species is
assigned to the universe (I mean according to shape
or mass), no definite species to the intellect, nor to the
affection.

LAO.   Thus, these two potencies of the soul are never,
and can never be, satisfied in their object, because
they pursue it infinitely.

LIB.   This would be so if the object were infinite through a negative privation of an end, whereas it is infinite because of a positive affirmation of an end, infinite and without limit.

LAO.   Therefore, you distinguish between two species of the infinite, one privative, which can tend toward something, for it is potency; just as darkness is infinite and ends when light appears; the other is perfective and is related to action and completion; just as light is infinite whose end would be darkness and privation. [11] Thus, the intellect conceives the light, the good and the beautiful as far as the horizon of its capacity is extended, and the soul drinks divine nectar and from the fount of eternal life as much as its own vessel permits; it is evident that the light is beyond the circumference of the soul's horizon, but the soul will always be able to penetrate it more and more; similarly, nectar is infinite and the source of living water is inexhaustible, so that the soul can become ever more and more intoxicated.

LIB.   Then, any imperfection in the object or a lack of satisfaction in the potency does not follow; but instead, the potency is seized by the object and beatifically absorbed by it. Thus the eyes make their imprint upon the heart, that is, upon the intelligence, and excite in the will an infinite torment of gentle love, in which the pain of not having the thing desired is absent, and present is the joy of ever finding the thing sought; and in the meantime satiety never arrives, because the appetite and consequently the taste never cease to desire. This is not the case with the nourishment taken by the body, which, after it has been filled up, loses the taste of the food so that it enjoys it neither before nor after indulging, but only at the moment of eating, and beyond a certain limit will feel nothing but discomfort and nausea.

You see, then, according to a certain similitude, how the highest good must be infinite, and how the impulse of the affection toward it must also be infinite, so that it will never cease to be a good—unlike the nourishment which is good for the body and becomes a poison when used immoderately. This is why the moisture of the ocean does not extinguish that flame,

---

[11] PLOTINUS *Enneads* 2. 4. 15.

and why the rigor of the Arctic Circle never tempers that ardor. That is why the heart is the captive of a hand which holds it and wants it not; holds it, because it belongs to it; wants it not, because, as though to flee from it, that hand escapes the more the heart aspires toward it; and the more the heart pursues it, the more it appears remote because of its most eminent excellence, according to the words, *Accedet homo ad cor altum, et exaltabitur Deus.* [12]

Such happiness of the affection begins in this life, and in this state has its own mode of being. Therefore one might say the heart is sheltered within the body and yet leaves it to be with the sun, meaning that the soul in the exercise of its two-fold faculty performs two functions, one of vivifying and activating a potentially animate body, the other of contemplating superior things; for just as the soul is in a receptive potency from what is superior to it, so is it in potential activity toward the body which is inferior to it. The body is as though dead and privative for the soul, which is its life and perfection; and the soul is as though dead and privative for the illuminating intelligence whereby the human intellect receives its proper character and actual form. For that reason the heart is said to be the principle of life and yet dead; to belong to a living soul when that soul does not belong to it. Because the heart is enflamed by the divine love, it is finally converted to fire and can enkindle whatever comes in contact with it; for having contracted the divinity to itself it becomes god-like, and consequently its aspect has the power to inspire love, just as in the moon the splendor of the sun can be contemplated and glorified.

And now for that which pertains to a consideration of the eyes, note that the present discourse attributes two functions to them, one of impressing the heart, [13]

---

[12] Ps. 63:7:

...Man shall come to a deep heart,
and God shall be exalted...

[13] PETRARCH *Rime* 195:

...od ella sani 'l colpo,
ch'Amor coi suoi belli occhi al cor m'impresse.

...or she may cure the blow,
that love with her beautiful eyes impressed
upon my heart.

the other of receiving an impression from the heart. Similarly the heart has two functions, one of receiving an impression from the eyes, and the other of making its impression upon them. The eyes apprehend the species and propose them to the heart; and the heart desires them and transmits its desire to the eyes; these conceive the light, diffuse it and enkindle the fire in the heart; the heart, burned and inflamed, sends its humour on the way to the eyes so that they may digest it. [14] Thus in the first place the cognition moves the affection which in turn moves the cognition. When the eyes act as stimulants they are cold, for they function as mirrors and transmitters of images; but when they are themselves moved, they are turbulent and altered, and they act as zealous performers, inasmuch as at first the speculative intellect sees the beautiful and the good, then the will longs for it, and in turn the diligent intellect becomes anxious about it, pursues and seeks it. The weeping eyes symbolize the difficult separation of the thing desired from him who desires it, which, because it does not satiate or weary him, offers itself as an infinite effort, and therefore is always with him and is something for which he never stops searching. Similarly, the felicity of the gods is described by their drinking of nectar and not by their having drunk it, by their tasting and not by their having tasted ambrosia, by their ceaselessly desiring food and drink and not by their having been gorged so that they have no desire for them. Therefore, the gods

---

[14] ANDREAS CAPELLANUS, *De amore*, trans. John J. Parry (New York, 1941), pp. 28-30; see also DANTE *Vita nuova* 20:

> Amor e cor gentil sono una cosa,
> Siccom, il Saggio in suo dittato pone;
> ... ...
> Beltate appare in saggia donna pui,
> Che piace agli occhi sì, che dentro al core
> Nasce un desio della cosa piacente.

> Love and the gentle heart are one same thing,
> even as the wise man teaches in his rime;
> ... ...
> Beauty is seen in wise lady then,
> which is so pleasing to the eyes that within the
> heart a desire is born for the pleasing thing.

For William Shakespeare's analogous allusion to the conflict between the eyes and the heart see *Love's Labour's Lost*, II, i, 228-242.

hold satiety to be a state of movement and apprehension and not a state of repose and comprehension; their satiety is never without appetite, nor do they experience appetite without being in some way satiated.

LAO.   *Esuries satiata, satietas esuriens.* [15]

LIB.   Precisely that.

LAO.   From this I can now understand how it has been said without reproach but with much intelligence and truth that the divine love weeps in inexpressible groans, for possessing all, it loves all, and loving all, it possesses all.

LIB.   But many a gloss would be necessary in order to make us understand the divine love which is the deity itself; whereas it is easy to understand divine love as it manifests itself in its effects and in inferior nature; I do not speak of the love that is diffused from the divinity among things, but of that love which from things aspires to the divinity.

LAO.   We shall have every leisure to return to this and other subjects. Let us depart.

END OF THE THIRD DIALOGUE

---

[15] "A satiated hunger, and a hungry satiety."

FOURTH DIALOGUE

INTERLOCUTORS

SEVERINO   MINUTOLO [1]

SEV.   Let us hear the discourses of nine blind men, who
            give nine reasons and particular causes of their blind-
            ness, although all of them agree that the general cause
            is the frenzy they have in common.
MIN.   Start with the first one.
SEV.   Although the first one is blind by nature, he none
            the less utters a love complaint and tells the others
            he cannot persuade himself that nature has been more
            uncivil to them than it has been to him; for even
            though they no longer see, they nevertheless have
            once experienced sight and have experienced the
            dignity of the sense and the excellence of sensible
            things which caused them to become blind; but he
            has come into the world like a mole, to be seen while
            he himself does not see and to long for things he has
            never seen.
MIN.   Many are found smitten by love, if we credit the
            rumor. [2]

---

[1] According to the *Fuochi di Nola* Francesco Severino was one of
three sons of Giambattista Severino of Nola. Francesco served as a soldier
under the command of the Marchesa of Misuraca, and later under Camillo
Pignatello. He was, Michel tells us, a comrade in arms of Bruno's father.
    Minutolo was probably Giovan Geronimo Minutolo, who lived with his
wife, three children and two servants in the hamlet, Livardi, near Nola. See
Fuochi di Nola, 1563, fol. 67 r°. See also SPAMPANATO, *Vita*, p. 27 and
note 3, and pp. 37, 53, 64; and BRUNO, *Des fureurs héroïques*, Paul-Henri
Michel, p. 454.
    [2] See *Vinculis in genere*, xxiii (*Caecitas vincientis*) in *Opera*, III, 665:
"Occulta etiam maxima ex parte etiam sapientibus vinculorum est ratio."

SEV.   He says that they at least have the happiness of retaining that divine image in their mind's eye, so that, no matter how blind they are, they nevertheless maintain within their fantasy that which for him it is impossible to have. Then in the sestet he turns to his guide and begs to be led to some precipice so that he may no longer be a horrid spectacle of nature's disdain.

Listen to his plea.

The First Blind Man Speaks

Oh happy ones who at one time have been able to see, though now you weep for the lost light, my companions, you once knew the two illuminations. For me these were neither enkindled, nor extinguished.

Thus a heavier misfortune than you believe is mine, and is worthy of greater lamentation. Nothing convinces me that nature has been more harsh with you than with me.

Oh guide, if you wish to bring me content, lead me to the precipice, so that my torment find a remedy. [3]

To be seen and yet not to see the light, like a mole I came forth into the world to be a useless burden to the earth.

The next one follows, who, bitten by the serpent of jealousy, has become infected in the visual organs. He goes without any guide, unless we may call jealousy the only guide he has. Because there is no remedy for his misfortune, he begs one of those around him to pity him and make him lose all sense of his

---

[3] Bruno's nine blind men follow the lament of the three blind men in ANTONIO EPICURO's *Dialogo di tre ciechi*. Epicuro's three lovers have lost sight of the vision of ideal beauty, and wish their own death. They regain their vision when a priest of Love leads them to Love's temple to invoke mercy of their Ladies. See BRUNO, in *Opere*, II, 486-487 and note 1.

evil by burying him with it, thus making him so hidden from himself, as the light of his eyes is now hidden from him. Then he says:

THE SECOND BLIND MAN SPEAKS

From her terrible tresses Alecto has torn
the infernal serpent, whose fierce bite has so
cruelly infected my spirit, that of my senses,
the most noble has perished,

depriving my intellect of its guide. That
mad rage of jealousy makes me stumble so on
every path, that in vain does my soul ask anyone for aid.

If no magic chant, or sacred herb, or
virtue of precious stone, or divine aid offer
me release,

may one of you, in the name of God, be
so merciful as to remove me from my own sight
by burying me without delay with my misfortune.

Next, one follows who says he has become blind by having unexpectedly emerged from the darkness into a great light; for accustomed to contemplating ordinary beauties, suddenly he was presented with one celestial beauty, a divine sun. As a result his sight was destroyed and extinguished was the twofold light which illumines the prow of his soul (for the eyes are like two light-houses guiding the ship); and his fate was similar to that of one who, nurtured in Cimmerian obscurity, suddenly fixed his eyes upon the sun. [4] And in the sestet he begs that he may be given passage to the inferno, because darkness only is suitable for so dark a being.

---

[4] PLATO *Republic* vii, 515.

THE THIRD BLIND MAN SPEAKS

If the sun suddenly appears to a man
nourished in profound darkness or under the
sky of the Cimmerian people, where the great
star diffuses a distant glow,

this inimical sun extinguishes the two-
fold light resplendent at the prow of the
soul and renders itself invisible. So was
my sight extinguished, for it was accustomed
to gazing upon vulgar beauties. [5]

Let me descend into hell! Why do I, a
dead man, go wandering through the world?
Why do I, an infernal clog, among you who are
living

go mingling with others? Why do I taste
the air in pain? Why am I put to so many
pains for having seen the supreme good?

The fourth blind man in his turn exposes the reason for his blindness, a reason similar, though not
identical with the preceding one. This blind man did
not suddenly find himself beneath the ray of light;
it is for having gazed upon it too often or for having
fixed his eyes upon it too much, that he has ceased
to be aware of any other light; thus one cannot say
that the ray of that unique light was the cause of his
blindness. And he says the same thing happened to
his sense of sight that happened to his sense of hear-
ing; for they who have accustomed their ears to great
uproars do not hear minor noises, as in the famous
example of the people of Cataduppia, who live where
the great Nile river descends precipitously from a
very high mountain upon the plain below.

MIN.   Therefore all those who have accustomed their
body and soul to the most difficult and the greatest
things, usually do not concern themselves with minor
difficulties. And this one ought not to be unhappy
because of his blindness.

---

[5] ANTONIO EPICURO, *La Cecaria*, ed. Palmarini (Venice, 1525), p. 63.

SEV. No, indeed. And he is called willingly blind, since he prefers that all other objects be hidden from him, for they could only annoy him by turning his view from that object alone which he desires to contemplate.

And in the meantime, he begs the wayfarers to aid in preventing him from falling upon some evil fortune, as he goes forth intent and wholly captivated by his chief object.

MIN. Refer us to his words.

SEV. THE FOURTH BLIND MAN SPEAKS

Falling precipitously from its height the Nile has abolished the sense of every other sound for the hapless Cataduppian people. So do I remain with spirit all intent

upon the most living light which illumines the world, and I am insensible toward all lesser splendors; and while this light shines upon the world, it willingly pays attention to no others.

I beg of you, warn me of running against some stone, or wild beast, and (tell me)[6] if I must descend or ascend,

so that these wretched bones may not fall into some open ditch, while I make my way deprived of guidance.[7]

It befalls the blind man who follows that because of the excessive weeping which has darkened his eyes, he cannot extend their visual rays to the visible species, and above all to that light again which, inspite of himself and at the cost of his great pain, he once saw. Moreover, he does not deem that his blindness is any

---

[6] Parentheses mine.

[7] The lament of the fourth blind man for lack of a guide follows a similar complaint of two of Epicuro's three blind men, who are in an analogous predicament. See EPICURO, *Dialogo di tre ciechi,* Palmieri (Venice, 1525), p. 42.

longer a passing disposition, but habitual, and privative
in the highest degree; for the luminous flame which
enkindles the soul through the pupil of the eye has been
too long and too vigorously repressed and oppressed
by a contrary humour; so that, no matter how much
he may cease from weeping, he is not persuaded that
the desired sight will be given him. And hear what
he says to his companions, so that they might give
him free passage.

THE FIFTH BLIND MAN SPEAKS

Eyes of mine, forever so pregnant of water,
when will the spark of your visual ray be thrust
forth over so many and so dense obstacles,

that I may see those sacred lights again,
the sources of my sweet pain? But ah! I believe that visual ray is forever extinct, so
long has it been oppressed and vanquished by
its contrary humour.

Let this blind one pass, and turn your eyes
to these founts, which overcome all other rivers
combined in one.

And if there is anyone who dares to dispute
it with me, I have reason to render it certain
that my two eyes contain an ocean!

The sixth blind man is in darkness, because by
excessive weeping he has poured forth so many tears
that all the moisture in him has been dried up, even
to the humid crystal of the eye, the diaphanous body
traversed by the visual ray which had formerly introduced the external light and visible species; from that
moment his heart was so afflicted that all the humid
substance (whose function it is to maintain the unity of
his diverse and contrary elements) [8] was consumed in

---

[8] In Bruno's view, it is water that endows the particles of the human
organism and of earth with cohesion. See GIORDANO BRUNO, De l'infinito,
universo e mondi, in Opere italiane, I, 359-361.

him; and love's affection remained in him without causing any tears, because his organism was dissolved by the victory of the other elements; as a result, he lost his sight and at the same time the cohesion of the parts of his body. Listen to the complaint he addresses to those around him:

THE SIXTH BLIND MAN SPEAKS

Eyes not eyes; [9] fountains no longer, you
have poured out all the moisture which holds
the body, the spirit and the soul together.
And you, crystal of the eye, which made

so many external objects known to the soul,
even you are consumed by my afflicted heart.
Therefore, arid and blind I lead my steps toward
the dark infernal cavern.

Ah, do not be niggardly in your mercy toward me, make me go promptly; I who in those
dark days took pleasure only in my tears

and was the source of so many streams;
now that every humour in me is dried up, toward
profound oblivion give me passage.

The next blind one has lost his sight from the intense flame which, issuing from his heart, has first consumed his eyes, then licked up all the remaining moisture of his body, so that, reduced to ashes, the lover is no longer himself; for the fire, whose virtue dissolves bodies into their atoms, has converted him into dust—an irremediable desegregation, inasmuch as water alone reassembles and combines the atoms of other bodies to make one subsistent composite. Nevertheless, he continues to experience the most intense fire. For that reason in the sestet he asks that a large passage be opened for him, for if anyone should be touched by his flame, he would become so insensible of the infernal fires, that he would no longer distinguish heat from cold snow. Therefore he says:

---

[9] PETRARCH *Rime* 161: "O occhi miei, occhi non già ma fonti..."

THE SEVENTH BLIND MAN SPEAKS

Beauty, rushing from my eyes to the heart,
formed in my breast a high furnace which, sending its relentless flame to the sky, absorbed
the moisture of my eyes;

then to appease its ardor it devoured all
my body's liquid elements, so that I should remain ever disjoined and reduced to separate
atoms of dust.

If you have horror of an infinite evil,
stay away from me, oh people! Beware of my
scorching flame, for if the contagion of its
fire assails you, you would seek winter in
hell's flames.

The eighth blind one follows, whose blindness was
caused by the arrow Love sent through his eyes to
penetrate his heart. As a result, he complains not only
of being blind, but also of being wounded, and more
profoundly burned than he believes any one could be.
His meaning is understood without difficulty in this
poem:

THE EIGHTH BLIND MAN SPEAKS

Vile assault, cruel blow, unjust palm,
acute point, devouring bait, strong sinew,
bitter wound, pitiless ardor, harsh burden,
arrow, fire and noose of that insolent god,

who pierced my eyes, burned my heart,
bound my soul and made me blind at one stroke,
a lover and a slave, so that in my deep blindness
every moment, everywhere and in every way I feel
my wound, my fire and my noose. [10]

---

[10] PETRARCH *Rime* 133; see also above, "Argument of the Nolan", p. 72.

Men, heroes and gods who inhabit the earth,
the inferno or Olympus, tell me, I beg you, how,
when and where

have you, among the oppressed, the damned—
among lovers, ever experienced, seen or heard
those who give vent to such complaints and to
so many of them?

The last blind one finally approaches, and he is also
mute; for, lacking the boldness to say the thing he
most desires without giving offense or invoking scorn,
he is unable to say anything at all. He is silent, but
he who guides him speaks in his place. Because his
discourse is without difficulty, I shall not comment on
it, but simply report it.

THE NINTH BLIND MAN'S GUIDE SPEAKS

You other blind lovers are fortunate, for
you can explain the reason for your blindness.
And the virtue of your tears can win you the
favor of gracious and chaste acceptance.

But the blind man I guide, torn with desire more than all the others, keeps his flame
hidden, mute perhaps for lack of boldness to
make clear his torment to his goddess.

And you, oh people unaware of these sad
obstacles, have compassion for this face become extinct, provide a path

for this afflicted body, consumed by
fatigue, which goes knocking at the door of
a less painful and more profound death.

Thus nine reasons have been indicated why the
human intelligence is blind with regard to the divine
object upon which it is unable to fix its eyes.

Of these reasons, *the first* personified by the first
blind man, is that the nature of our species, according
to the rank in which it finds itself, always aspires higher
than it can attain.

MIN.   Because no natural desire is vain, we may be sure that there is outside the body a more excellent state to which the soul can be united when it is raised nearer to its object.

SEV.   As you point out very well, no natural potency or impulse is without its reason for being, which is, in fact, the rule of nature which orders things. Therefore it is absolutely true for every well disposed mind that the human soul (such as it appears while residing in the body) shows by everything it expresses that it is a stranger in this country, for it aspires to the universal truth and good, and is not satisfied with what is offered to it for the use and profit of its natural species.

The second reason, personified by the second blind man, proceeds from the disturbance of the affection which, when one is in love, is jealousy, and jealousy is like a worm for whom the same subject is enemy and progenitor, for it nibbles at the cloth or wood from which it is generated.

MIN.   It seems to me that such jealousy has no place in heroic love.

SEV.   No, not for the same reason it is found in vulgar love; but I understand jealousy in a different though corresponding way, according as it is manifest among lovers of the true and the good when they are incensed against those who would adulterate, waste, or corrupt the true and the good, or in one way or another treat them with indignity. And they are incensed against them to such an extent, that, should they fall into the hands of those men, they are tormented, done to death, and treated ignominiously by the ignorant populace and vulgar sects.

MIN.   Certainly, no one sincerely loves the true and the good without becoming irate against the multitude, just as no one experiences vulgar love without being jealous and fearful for the thing loved.

SEV.   And thus he will be truly blind to many things, and according to the common opinion, stupid and mad in the highest degree.

MIN.   I have noted a passage [11] which says that all those are stupid and mad who have any sense beyond and above the universal sense of ordinary men. But this madness is of two kinds, accordingly as some surpass

---

[11] MARSILIO FICINO *In convivium* vii, 12-14.

or mount above the limit to which all or a majority of men ascend or can ascend (such men are thus inspired by the divine frenzy), or as some descend lower, falling to the level of those who lack sense and reason, and lack them more than the multitude of ordinary men. This last species of madness, lunacy and blindness will not attain heroic jealousy.

SEV. *The third* reason, personified by the third blind man, proceeds from this, that the divine truth, in the mode of the supernatural, called metaphysics, is revealed to the rare spirits whom it favors, and does not submit its arrival to measurements of movement and of time, as is the case in the physical sciences (those acquired by the light of nature which proceed from a thing known by sense and reason to a thing still unknown, in the discursive mode one calls argumentation), but, on the contrary, arrives suddenly and unexpectedly according to the mode appropriate to its activity. For that reason the sage said, *Attenuati sunt oculi mei suspicientes in excelsum.* [12] Therefore, a vain length of time, laborious study and effort of research are not required for obtaining divine truth, but it allows itself to be absorbed as promptly as the light of the sun renders itself present to him who turns and opens himself to it. [13]

MIN. Would you say then, that scholars and philosophers are not more apt to receive this light than the ignorant are?

SEV. That might be true in one sense, and might not be true in another. It does not make any difference when the divine spirit, by its own providence, communicates itself without any special disposition of the subject who receives it; that is, when it communicates itself because it seeks out and elects the subject of its own accord. But it makes a great difference when the divine spirit waits and wishes to be sought, and then at its good pleasure would be discovered. In this mode it does not appear to everyone, nor can it appear to anyone unless

---

[12] Isa. 38:14:

...My eyes are weakened as they gaze into the heavens...

[13] See PLOTINUS *Enneads* 5, 3, 17. See also FELICE TOCCO, *Le opere latine di Giordano Bruno* (Firenze, 1889), pp. 369-370.

he seeks it. And so it is said, *Qui quaerunt me invenient me;* [14] and elsewhere, *Qui sitit, veniat et bibat.* [15]

MIN.

It cannot be denied that the apprehension of the second mode comes with time.

SEV.

You are not distinguishing between disposing oneself to the divine light and apprehending it. Certainly I do not deny that in order to dispose oneself to it, time, discourse, zeal and labor are required; but, alteration, as we say, comes with time, and generation, in an instant; or further, as anyone can see, it takes time to open a window, but the sun enters in a moment. The same thing applies to what we have been saying.

The fourth reason, personified by the fourth blind man, is entirely without the indignity belonging to the habit of sharing the errors of the mob—errors which can either be far removed from all philosophical opinion, or derived from the study of vulgar philosophies esteemed true by the mob the more they conform to the mob's view. This is one of the greatest and most unseemly habits into which one can fall; for as Al-Gazeli and Averroës have shown us by examples, there are those who from infancy and youth have accustomed themselves to digesting poisons, so that in the long run these poisons have become to them sweet and appropriate nourishment for their organisms, while they hold in abomination things truly sweet and good for normal beings. [16] The blindness of the fourth lover has a most worthy reason, for it comes from the habit of gazing upon the true light (a habit which, as it has been said, cannot be practiced by the many). This blindness is heroic and appropriate for the worthy satisfaction of our blind lover, who, far from finding

---

[14] Luke, 11:9-10:

...Ask and it shall be given you: seek and you shall find: knock and it shall be opened to you.

See also Matthew 7:8.

[15] John, 7:37:

...If any man thirst, let him come to me and drink.

[16] For an analogous statement of this idea see *Theses de magia*, xxxi, in *Opera* III, 475.

any remedy for it, truly arrives at the point of scorning every other sight, and asks nothing of the human community but free passage and progress toward contemplation, because too frequently he is a victim of snares and is usually jostled against mortal obstacles.

The fifth reason, personified by the fifth blind man, proceeds from the lack of proportion between the means of our intellect and the intelligible object; for to contemplate divine things we must consider them by means of symbols, similitudes and other ambiguities which the Peripatetics call phantasms; moreover, we must proceed by the agency of the creature to the speculation of its essence, by the way of the effect to the notion of cause; all means so inadequate for attaining such an end, that they would seem rather to be obstacles, if one must believe that the highest and most profound knowledge of divine things is negative and not affirmative, knowing that the divine beauty and goodness is not something which can fall and submit itself to our concept, but something completely beyond our comprehension, especially in this mortal state, called by the philosopher a speculation of phantasms, and by the theologian, a vision only by similitude, mirror and enigma. [17] For we do not truly see the effects and the true forms of things, or the substances of ideas, but we see only the shadows, vestiges and images of them, for we are like those who are inside the cave and from birth turn their backs to the light and their faces to the dark, so that they never see that which truly is, but the shadows of those things whose substance is to be found outside the cave. [18]

That is why a spirit comparable to Plato, if not superior, weeps for the clear vision he has lost, and desires to exit from the cave, in order to see his light again not by reflection, but by an immediate conversion. [19]

MIN.   What this blind man deplores, it seems to me, is not the difficulty caused by the reflected vision, but

---

[17] Cor. 13:12:

> We see now through a glass in a dark manner;
> but then face to face. Now I know in part;
> but then I shall know even as I am known.

[18] PLATO Republic vii, 515-517.
[19] PLOTINUS Enneads 4. 8. 1. See also TOCCO, Opere latine di Giordano Bruno, p. 369 and note 4.

the difficulty caused by the intermediary interposed between his visible potency and the object.

SEV.

Although these two modes are distinct in the sensitive cognition or the sensitive sight, they suddenly concur in one rational or intellective cognition.

MIN.

I believe I have read and understood that every vision requires an intermediary between its potency and the object. For, just as by means of light diffused in the air, and by the image of an object which proceeds in some way from the thing seen to him who sees it, the act of vision becomes effective, so in the intellectual sphere where the sun of the active intellect shines, by means of the intelligible species which receives its form from the object, and so to speak, proceeds from it, our intellect or some other inferior one begins to comprehend something of the divinity. For, just as our eye, when we see, does not receive the light of fire or of gold in substance, but in similitude, so our intellect, in whatever state it is found, does not receive the divinity in substance (for then there would be as many gods as there are separate intelligences), but receives it in similitude; and this is why these intelligences are not formally gods, but may be designated divine things, the divinity and the divine beauty remaining one and exalted above them all.

SEV.

You explain it very well; but this explanation does not oblige me to retract anything, for I have not said the contrary. It is necessary only that I explain myself. Thus first I declare that the immediate vision about which we have spoken and have understood each other does not exclude those intermediaries such as the intelligible species or the light, but excludes rather those which correspond to the thickness and density of a diaphanous mean or even to the opacity of a body interposed, as it happens to him who looks through more or less turbid water, or cloudy and murky air, that he would desire to see without an intermediary, if permitted to gaze through pure, lucid and clear air. All of which you have more or less explained by the words, *thrust forth over so many dense obstacles*. But let us return to our discourse.

The sixth reason, personified by the sixth blind man, is none other than the weakness and inconsistency of the body which is in continual motion, change and alteration, and where operations must conform to the aptitudes resulting from the condition of its nature and being. For how would you have immobility, persistence, entity and truth belong to a thing which changes every moment from one thing to another, and is ever in the process of becoming something else? What reality, what image can be retained, depicted and impressed upon the eye, when the pupils are dispersed in water, when the water turns into vapor, vapor into flame, the flame into air, and so on, while a sensible and knowing subject endlessly perambulates the wheel of metamorphoses?

MIN.   The movement is one of alteration; he who is moved is always another, and he who is another always bears himself and behaves otherwise than he did before, for intellection and affection conform to the reason and the condition of the subject. And he who is always ·another, who forever changes his vision, can only be completely blind with respect to the beauty which is always unique and one, which is unity itself, entity and identity.

SEV.   Exactly.

The seventh reason, allegorically contained in the complaint of the seventh blind man, derives from the fire of the affection, from which some become impotent and incapable of apprehending the truth, inasmuch as their affection overcomes their intellect. Such are those who place love before understanding, so that everything appears to them colored by their affection; for it is an established fact that for those who would attain the truth by way of contemplation a perfect purification of the thought is necessary.

MIN.   We know very well that there is a great diversity among those who contemplate and those who seek. Some (following the habits of primary and elementary disciplines) advance by way of numbers, others progress by way of figures; some advance by the rules or without the rules, others progress by way of composition and division; some by way of separating into parts and assembling them again, others by inquiry and disputation; some by discourse and definition, others by the interpretation and deciphering of terms, vocabularies and dialects; in other words, some are mathematical philosophers, and others are metaphysicians, logicians, or grammarians. The same diversity exists among those for whom to contemplate is to study written opinions and to apply their attention to them;

so that it comes to this that the same light of truth expressed in the same book and by the same words could serve the designs of numerous sects, diverse and hostile among themselves.

SEV.   That is why the affections have such power to impede the apprehension of the truth, inasmuch as those who submit to them are incapable of perceiving it, as those who attribute to the food the bitterness of their mouth submit to the malady of stupidity.

Now such a species of blindness is noted in this blind man, whose eyes are altered and deprived of their natural power by that which has been sent from the heart and impressed upon them, altering not only their sight, but all the other faculties of the soul besides, as the present allegory demonstrates.

With regard to the meaning of the *eighth* blind man, as he has lost his sense of sight by the impact of a sensible object, so has his intellect been blinded by the excellence of the intelligible object. Thus it happens that he who sees Jove in his majesty loses his life, and consequently loses his sense. So does it occur that he who so gazes on high sometimes becomes overwhelmed by majesty. [20] Besides, when he would penetrate the divine species, it pierces him like an arrow.

Therefore, the theologians say that the divine word is more penetrating than the point of a sword or knife. [21] Wherever it forms and impresses its image, no other form can be impressed or sealed; for where such an impression has been made, a new mark cannot replace it without the first one having yielded; consequently it may be said that a being no longer has the faculty of receiving another form, even if there is anyone who attempts to change or transform it through a necessary alteration of proportion.

The ninth reason is personified by the ninth man who is blind because of lack of confidence and humility

---

[20] Prov. 25:27:

As it is not good for a man to eat much honey, so he that is a searcher of majesty, shall be overwhelmed by glory.

[21] Hebrews, 4:12:

For the word of God is living and effectual, and more piercing than any two-edged sword; and reaches into the division of the soul and the spirit, of the joints also and the marrow, and is a discerner of the thoughts and intents of the heart.

of spirit, both of which are caused by great love, for he fears his ardor may give offense. With reference to which the Canticle says, *Averte oculos tuos a me quia ipsi me avolare fecere.* [22] And, therefore, he curbs his eyes from seeing what he most would desire and enjoy, as he holds his tongue from speaking to whom he most longs to speak, for fear that some defect of his glance or of his word might debase him, or in some way cause him disgrace. And this is what happens when the excellence of the object is so far superior to the power of apprehension. For this reason the more profound and divine theologians say God is honored and adored more by silence than by words, and that to see him better one must close one's eyes to the species represented than open them. This is why the negative theology of Pythagoras and Dionysius is so highly renowned above the demonstrative theology of Aristotle and the schoolmen. [23]

MIN.   Let us depart and discourse on the way home.

SEV.   As you like.

END OF THE FOURTH DIALOGUE

---

[22] Cant. 6:4:

Turn away thine eyes from me, for they have made me flee away. Thy hair is as a flock of goats, that appear from Galaad.

[23] BRUNO, *Cabala*, in *Opere*, II, 269:

...Fidele colui che non permette che siano tentati sopra quel che possono: lui conosce li suoi, lui tiene e mantiene gli suoi per suoi, e non gli possono esser tolti. O santa ignoranza, o divina pazzia, o sopraumana asinità. Quel rapto, profondo e contemplativo Areopagita, scrivendo a Caio, afferma che la ignoranza è una perfettissima scienza; come per l'equivalente volesse dire che l'asinità è una divinità. Il dotto Agostino, molto inebriato di questo divino nettare, nelli suoi *Soliloquii* testifica che la ignoranza più tosto che la scienza ne conduce a Dio, e la scienza più tosto che l'ignoranza ne mette in perdizione...

...Faithful is he who does not allow himself to be tempted beyond what his is capable: this one understands what belongs to him, and he holds and maintains his own for himself, and his own cannot be taken from him. Oh holy ignorance, oh divine madness, oh superhuman stupidity! That enraptured, profound and contemplative Areopagite, writing to Caius, affirms that ignorance is a most perfect knowledge; as one would say equally that stupidity is a divinity. The learned Augustine, much inebriated of this divine nectar, in his *Soliloquies*, testifies that ignorance more than knowledge leads us to God, and knowledge more than ignorance brings us to perdition...

LAODOMIA   GIULIA [1]

LAOD.   Some other time, oh sister, you will understand
the significance of the complete story of these nine
blind men. They were nine most handsome and loving
youths and so ardently smitten by the graciousness of
your sight, that, having lost hope of gathering love's
longed for fruition, and fearing that such despair would
reduce them to ultimate ruin, they departed from the
happy Campanian fields; and (they who were rivals)
commonly agreed to swear by your beauty never to
separate until they had tried everything to find one
more beautiful than you, or at least, one similar to you
and, besides, adorned with that mercy and pity of
which your cruel heart was destitute; for they believed
this was the only remedy that could release them from
their cruel captivity. On the third day after their
departure, as they passed not far from the mount of
Circe, it pleased them to go and see those antique

---

[1] Both ladies, according to the archives of Naples, were of the family
of Savolino and were maternal cousins of Giordano Bruno. Laodomia was
born 1550 and Giulia, 1544. Spampanato goes so far as to suggest that
Laodomia was for Bruno a Beatrice whose favor raised him to the divine
contemplation alluded to in the sonnet to Diana in the second dialogue of
Part II. However, Bruno concludes the last dialogue of *De gli eroici furori*
with Giulia's admission that her obduracy, and not Laodomia's, was ultimately the means by which her lovers were raised to grace. See *Fuochi*,
1545, fol. 97 v°, 1563, fol. 56 r°; see also SPAMPANATO, *Vita*, p. 64 and
note 3, *Postille*, p. 235-236, BRUNO, in *Opere*, II, 507 and note 2, and
BRUNO, *Des fureurs héroïques*, Michel, p. 454-455.

caves and sight consecrated to that goddess. When they arrived there, because of the majesty of that solitary and windy place, and the majesty of the high and resounding rocks, and of the murmuring sea waves which broke into those caves, [2] and owing to other circumstances which the place and season offered, all of them became as though inspired and one among them (who it was I shall tell you), more impassioned than the others, spoke these words: "Oh would that heaven would be pleased to present us at this time, as happened in other happier centuries, with that magician, Circe, who by virtue of plants, minerals, venoms and incantations was able to seize control of nature. Implacable as she may be, I firmly believe that she would be merciful to us in our misfortune. Solicited by our supplication and complaints, she would condescend to provide us with a remedy and to accord us the favor of vengeance against our cruel enemy." Hardly had he finished speaking these words, when suddenly before the eyes of everyone, a palace appeared which anyone with any notion of human accomplishment could easily see was no work of man or nature, whose aspect I shall describe at another time. Stricken by that great marvel and moved by hope that some propitious deity (the cause of this apparition) would explain the state of their fortune, they cried out together that nothing could befall them worse than death, which they deemed less evil than to go on living in such intense suffering. This is why, not finding the door closed to them or any porter who inquired what their business was, they entered, and found themselves in a most rich and ornate room, where, in that regal majesty in which Apollo was discovered by Phaeton, appeared she who is called his daughter, [3] at whose appearance they saw disappear the images of many other deities who used to minister to her. Received and encouraged by her gracious visage, they advanced, and overcome by the splendor of that majesty they fell upon their knees, and all together in varied strains dictated by their diverse

---

[2] This description may refer to one of those vistas which Bruno saw on his way to Rome, or at Gaeta. See BRUNO, *Opere italiane,* ed. Gentile, II, 508 and n. 2.

[3] VIRGIL in the *Aeneid,* vii, ii, alludes to Circe as daughter of the sun; see also OVID *Metamorphoses* ii, 19-46.

talents, offered prayers to the goddess. To conclude, they were treated by her in such a way, that blind, wandering and miserably belabored, they traversed all the seas, passed every river, overcame every mount, traversed every plain for a period of ten years, [4] after which beneath the temperate sky of the island of Britain, they found themselves in the presence of the lovely and gracious nymphs of Father Thames. [5] After they had performed acts of appropriate humility, which were received with gestures of the most chaste courtesy, one among them, their chief, whose name I shall give you another time, expressed the common cause in a tragic and lamenting tone as follows:

> Noble ladies, the bearers of a closed ves-
> sel present themselves before you, their hearts
> pierced through, not by an error of nature, but
> by a cruel fate which tortured them with this
> living death, and they remain in blindness.

> We are nine spirits who, wandering for
> many years because of the desire to understand,
> have travelled many countries, and we were one
> day victims of a severe and sudden disaster,
> which, if you listen to our story, will cause
> you to say, O worthy ones, and unhappy lovers!

> A cruel Circe, who boasts of having this
> beautiful sun her progenitor, received us after
> a long and adventurous voyage; she opened a ves-
> sel and sprinkled us with water, and to that
> gesture joined her incantation.

> Awaiting the consummation of such action,
> we were in silence and mute attention, until she
> spoke:—O, you sorrowing ones, depart, blind as
> you are in all things; go gather the fruit that
> falls to those who direct their gaze too high.—

> Then suddenly the blind men—Daughter and
> mother of darkness and horror (we said with one

---

[4] For the statement that Bruno may have been referring to the period in his life when he left Naples in 1576 and wrote *De gli eroici furori* in 1585, see GENTILE, in *Opere*, 510 and note 1.

[5] See above, "Argument of the Nolan", p. 65.

voice) does it please you, then, to treat wretched
lovers so cruelly who submit themselves before you,
willing perhaps to consecrate their hearts to you?

But when the frenzy suddenly excited by so
strange a mishap was somewhat appeased, each
one collected himself, and as rage yielded to
pain, all implored mercy, mixing the following
words with their tears:

—Now, if it pleases you, oh noble enchantress,
that zeal for glory may pierce your heart, or
that your heart be anointed and soothed by the
waters of compassion, have pity upon us with your
remedies, and close the wound inflicted upon our
hearts.

If your lovely hand be pleased to aid us,
do not delay that some sad one of us may reach
death before your gesture give us the right to
say, a great torment was caused by her, but a
much greater consolation.

And she replied: —O curious spirits, take
this other fatal vessel which my hand is powerless to open; [6] and go far and wide on a pilgrim-
age through the world, seeking out all the numerous kingdoms,

for destiny wishes that this vase remain
closed until lofty wisdom and noble chastity and
beauty together apply their hands to it; all
other labors are fruitless to pour forth this
water.

But if it happens that those gracious hands
with this water besprinkle whoever approaches
them for a cure, you will be able to experience
divine virtue, for your cruel torment being
changed to remarkable joy, you will see the two
most beautiful stars in the world. [7]

---

[6] The vessel is the hermetic vase of the *Opus Alchemicae*, symbol of
the uterus in which is found the philosopher's stone. See PHILALETHES, in
Musaeum Hermeticum (Frankfort, 1678), p. 770.

[7] Goodness and Truth, as primary and secondary intelligence; see above,
pp. 41-42.

May none of you be saddened, no matter how
long so much of the firmament may be concealed
in profound darkness; for no pain is so great
that will render you worthy of so great a good.

For the prize to which your blindness leads
you, hold vile every other gain and esteem every
torture as so much joy, for the hope of contemplating these unique and rare graces will incline
you to scorn every other light.—

Alas! Too long have our limbs gone wandering through the whole terrestrial earth, so that
finally we have come to believe a sagacious
beast has filled our hearts with false hope by
its promises.

Henceforth (although we know it is late)
we perceive that this enchantress, for our greater
woe, strives to keep us in eternal expectation.
For she believes that no lady of so many virtues
can be seen beneath the cloak of heaven.

Now, although we know every hope vain, we
yield to our destiny and are content not to re-
treat from painful labours, and are content to
advance (though trembling and weary), without
ever halting our steps, and to suffer for as long
a time as life remains in us.

Lovely nymphs who sojourn on the verdant
shores of the gentle Thames, ah, in God's name,
lovely ones, hold it not beneath you, even if
it is in vain, to lend your white hands to disclose what our vase conceals.

Who knows? Perhaps on these shores where
one sees this torrent, with its nymphs, so rapid-
ly rising as it rewinds itself to its source,
heaven has destined that she whom we seek may be
found.

One of the nymphs took the vase in her hand, and
without essaying further, offered it to each one of the
others, but none could be found who dared to open it
first. But all of them by common agreement, after

merely looking at it, referred and proposed it in defer-
ence and reverence to only one among them; [8] who
seized it finally, not so much from a desire to dem-
onstrate her glory, but though pity and the desire
to bring succour to these hapless men; and although
uncertain, she clasped it in her hand, and almost
spontaneously, opened it herself. How would you have
me relate how great was the applause of the nymphs?
Do you imagine I can express the excessive joy of
the nine blind men, who, having heard that the vase
was opened, felt themselves sprinkled with the longed
for water, opened their eyes, saw the twin suns and
were overwhelmed by a two-fold felicity, that of having
recovered the light formerly lost and that of having
newly discovered the other light which alone could
show them the image of the supreme good on earth? [9]
How, I ask, would you have me express that happiness
and jubilance of voice, that thrill of spirit and body
which they themselves were incapable of expressing?
For a moment they appeared to be in frenzied intoxi-
cation; they thought they were dreaming and seemed
not to believe what they manifestly beheld. But when
the excess of that frenzy finally became somewhat
subdued, they took their places in a circle, where

The first sang and played the guitar in this tone

O rocks, O trenches, oh thorns, oh twigs,
oh stones, oh mountains, oh plains, oh valleys,
oh rivers, oh seas, how you reveal yourselves
gracious and sweet, for heaven has discovered
to us your mercy and your worth! Oh steps
spent for good fortune!

The second played and sang with his mandolin

Oh steps spent for good fortune, oh goddess
Circe, oh glorious afflictions! Oh, how the pains

---

[8] Diana, chief of the nymphs of the wilderness. See "Argument of the
Nolan", p. 76; see also above, pp. 224-226.

[9] Diana, the highest Monad and unity of the primary and secondary
intelligences, and of the formal and material principles in the infinite.

of so many months and years are so many divine graces, if this is our recompense after so much torment and misery!

The third played and sang with his lyre

After so much torment and misery, if this is the port prescribed by our tempests, there remains nothing else for us but to thank heaven for having placed before our eyes this veil, through which this light has been finally revealed.

The fourth sang with his viol

Through which this light has been finally revealed, blindness more worthy than any other sight, cares more sweet than any other pleasures; for to the most excellent light you have led us, making less worthy objects useless to the soul.

The fifth one sang with his Spanish timbrel

Making less worthy objects useless to the soul, nourishing a noble thought with hope, was one who spurred us toward that unique path, which showed us the most beautiful creation of God. In this way fate will show itself propitious.

The sixth one sang with his lute

Fate will show itself propitious in this way. For fate does not wish that good follow good, or pain be the presage of pain; but making the wheel turn, it raises, then it hurls down, as in mutability, the day gives itself to night.

The seventh sang with his Spanish harp

As in mutability, the day gives itself to night, when the great cloak of the nocturnal torches ob-
[scures

the flaming chariot of the sun, so he who governs by
eternal decree crushes the great and raises the hum-
[ble.

The eighth one with bow and viol

He crushes the great and raises the humble, who
sustains his infinite schemes, and by a rapid, moder-
[ate,
or slow rotation he distributes in the immense crea-
[tion
all that is hidden and all that remains seen.

The ninth with a three-stringed viol

Oh, may all that is hidden and all that remains
seen not deny, but confirm the incomparable end
[of
our labors, whose witnesses are the fields and
mountains, ponds, r i v e r s, seas, rocks, trenches,
[thorns,
twigs and stones. [10]

After each one in this form and in his turn, had played
his instrument and sung his sestet, they danced together
in a circle, and, playing in a most sweet accord to the
praise of the unique nymph, sang a song which I think
I shall remember well enough.

GIU.   Don't fail, I pray you, sister, to let me hear as
much as you may recall.

LAO.   SONG OF THE ILLUMINATED [11]

"I no longer envy, O Jove, your firmament", says
Father Ocean with raised brow, "for I have so much
joy in what my empire offers".

---

[10] The pattern of the sestets in this *canzone* is of circular structure
appropriate to the theme of the wheel of fate. The last verse of each sestet
forms the first verse of each succeeding sestet, and the last lines of the last
sestet are identical with the first lines of the *canzone*. Bruno bases his
verse upon the rules for the *capceudadas* rhymes of Guilhem Molinier, "Las
leyes d'amors", in *Monumens de la littérature romane*, ed. M. Gatin-Cer-
noult (Toulouse, 1841), I, 237.

[11] This final *canzone* follows EPICURO's *Dialogo di tre ciechi* when
the three blind lovers express through the priest the illumination their ladies

"How haughty you are!" Jove replies.
"What else do you have beside your wealth? Oh
lord of the senseless waters, why do you so in-
flate yourself with such foolish boldness?"

"You have", said the god of the waters,
"in your power the blazing heavens, where the
fiery zone is, in which you can see the eminent
chorus of your stars,

"and through them the whole world gazes
upon the sun. But, I say, even the sun shines
with less brightness than She who makes me the
most glorious god of the great creation of
worlds.

"And I hold in my vast bosom, among all the
others that nation where the happy Thames is
seen, which has the pleasing chorus of the most
beautiful nymphs.

"Among these I possess one who is unique
among all beautiful ones, who will make you a
lover of the sea more than of the sky, oh loud
thundering Jove, for your sun shines with less
splendor among the stars."

And Jove replies: "O, god of the tossing
seas, that any one be found more blessed than
I is not permitted by fate, but my treasures
and yours run their course together.

"The sun prevails among your nymphs through
this one, and by the force of eternal laws and
of the alternate abodes, she is valued as the
sun among my stars."

I believe I have reported it to you completely.

GIU.   You may be assured of it, for there is no lack of

---

have provided them. Bruno's song, however, allegorizes his philosophy.
Jove with his firmament represents the superessential intelligence, and Father
Ocean represents the realm of Nature. Both rule with equal power through
Diana, who is brightest sun, yet also queen of nymphs. Accordingly she
is Bruno's philosophical symbol of the highest unity-of-opposites. See BRUNO,
in Opere, II, 517 and note 2; see also above, pp. 43-44, 226.

perfection in their argument, nor lack of art in the perfection of the strophes. As for myself, if by heaven's grace I have achieved any beauty, I believe I have been granted even a greater grace and favor; for whatever my beauty may have been, it was in some way responsible for the discovery of that unique and divine beauty. I am thankful to the gods, for in my youth when I was so young that the flames of love could not enkindle my heart, my cruelty and intractability, though simple and innocent, was the occasion and means of according my lovers graces incomparably higher than they could otherwise have obtained whatever might have been my benevolence.

LAOD. With respect to the souls of those lovers, I assure you that, just as they are not ungrateful to their enchantress, Circe, for their dark blindness, calamitous labours, and their bitter afflictions which brought them to so great a good, so will they not be less appreciative of you.

GIU. This is my desire and hope.

END OF THE SECOND AND LAST PART

of

THE HEROIC FRENZIES

BIBLIOGRAPHY

EDITIONS OF "DE GLI EROICI FURORI"

BRUNO, GIORDANO. *Opere italiane.* Nuovamente ristampati; con note da Giovanni Gentile. 2 ed. riveduta e accresciuta. Bari: Gius, Laterza & Figli, 1923-1927.

————. *De gli eroici furori.* Edited by Francesco Flora. Torino: Unione Tipografico Editore Torinese, 1928.

————. *Des fureurs héroïques.* Translated by Paul-Henri Michel. Paris: Société d'Édition "Les Belles Lettres", 1954.

————. *The Heroic Enthusiasts.* Translated by L. Williams. Introduction compiled chiefly from David Levi's "Giordano Bruno e la religione del pensiero". London: G. Redway, 1887.

————. *Eroici furori, oder Zweigespräche vom Helden und Schwärmer, übsersetzt und erläutert von Dr. Ludwig Kuhlenbeck.* Leipzig: W. Friedrich, 1898.

PRIMARY SOURCES

ALCIATI, ANDREA. *Emblematum flumen abundans.* Edited by Henry Green. Facsimile of Lyons edition by Bonhomme 1551. Published for the Holbein Society by A. Brothers. London, 1871.

ALIGHIERI, DANTE. *Convivio.* Edited by Valentino Piccoli. Torino: Bologna, Zanichelli, 1925.

————. *La Vita nuova.* Edited by Michele Scherillo. Milano: Ulrico Hoepli, 1911.

AQUINAS, THOMAS. *Summa theologica.* Vol. I of *Opera omnia.* Edited by Petri Fiaccadori. New York: Mesurgia Publishers, 1948-1950.

ARISTOTLE. Loeb Classical Library. London: William Heinemann Ltd., 1926.

De anima.
Metaphysics.
Nichomachean Ethics.
Physics.

AVERROËS. *Cordubensis commentarium magnum in Aristotelis De anima libros;* recensuit F. Stuart Crawford. Cambridge, Massachusetts Medieval Academy of America, 1953. (The Medieval Academy of America. Publication No. 59. Corpus philosophorum medei aevi. Corpus commentarium Averroës in Aristotelem. Versionem Latinorum Vol. VI, 1.)

BOCCACCIO, GIOVANNI. *Genealogie deorum gentilium libri.* Edited by V. Romano. 2 vols. Bari: Laterza & Figli, 1951.

BRUNI NOLANI, JORDANI. *Opera latine conscripta.* Recensebat F. Fiorentino. 8 vols. Neapoli, 1879-1891.

BRUNO, GIORDANO. *De la causa, principio e uno.* Edited by A. Guzzo. Firenze, 1933.

CUSANUS, NICOLAUS. *De docta ignorantia,* in *Opera omnia.* Edited by Ernestus Hoffmann and Raymundus Klibansky. Lipsiae: Felix Meiner, 1832.

————. *Vision of God.* Translated by E. G. Salter. London: J. M. Dent & Sons, 1928.

DE BEZE, THÉODORE. *Les Vrais Portraits des hommes.* N. p., 1581.

DUPLESSIS, GEORGE. *Les Emblèmes d'Alciat.* Paris: Rapilly, 1884.

EPICURO, ANTONIO. *La Cecaria.* Edited by Palmieri. Venice, 1525.

FICINO, MARSILIO. *In convivium,* in *Opera Platonis.* Paruo et Badio, 1518.

————. *Sur le Banquet de Platon de l'Amour.* Raymond Marcel. Paris: Société d'Edition "Les Belles Lettres", 1956.

GIOVIO, PAOLO. *Ragionamento di Mons. Paolo Giovio sopra i motti, et disegni d'arme, et d'amore, che communemente chiamono imprese. Con un discorso di Girolamo Ruscelli, intorno allo stesso soggetto,* Venezia, 1566.

LUCRETIUS. *De rerum natura.* Translated by R. E. Latham. London: Penguin Books, Ltd, 1952.

————. *De rerum natura.* Loeb Classical Library. London: William Heinemann, Ltd., 1925.

OVID. *Metamorphoses.* Edited by F. J. Miller. 2 vols. Loeb Classical Library. London: William Heinemann, Ltd., 1951.

PETRARCH, FRANCIS. *L'Africa.* Edited by Nicola Festa. Firenze: Sansoni, 1926.

————. *Rime Sparse.* Edited by E. Chiorboli. Milano: Casa Editrice Trevisini, n. d.

————. *Secretum Francisci Petrarche de Florecia Poete.* Strasbourg: Adolph Rusch, 1473.

————. *Secretum.* Translated by William Draper. London: Chatto and Windus, 1911.

PLATO. Loeb Classical Library. London: William Heinemann, Ltd., 1925.

Cratylus.
Phaedrus.
Republic. 2 vols.
Symposium.

PLOTINUS. *Enneads.* Edited by Emile Brehier. 7 vols. Paris: Société d'Edition "Les Belles Lettres", 1954.

RONSARD, PIERRE DE. "L'Abrégé de l'art poetique", in *Oeuvres complètes,* Vol. VII. Edited by Blanchemain. Paris: Bibliotheque Elzevirienne, 1866.

SEARS, GEORGE. *Collection of Emblem Books of Andrea Alciati.*

TANSILLO, LUIGI. *Poesie liriche.* Edited by Fiorentino. Napoli: Morano, 1882.

SECONDARY' SOURCES

Books.

AZZOLINA, LIBORIO. *Il dolce stil nuovo.* Palermo: Alberto Reber, 1903.

BADALONI, NICOLA. *La filosofia di Giordano Bruno.* Edited by Parenti. Firenze, 1955.

BRUNO, GIORDANO. *Heroische Leidenschaften und individuelles Leben.* Edited by Ernesto Grassi. Bern: Francke, 1947.

CESAREO, ALFREDO. "Amor mi spira", in *Miscellanea de studi critici edita in onore di Arturo Graf*, pp. 515-543. Bergamo: Istituto italiano d'Arti grafiche, 1903.

CORSANO, ANTONIO. *Il pensiero di Giordano Bruno nel suo svolgimento storico.* Firenze: 1940.

FENU, EDOARDO. *Giordano Bruno.* Copywright by Morcelliana. 1938 — xvi.

GENTILE, GIOVANNI. *Giordano Bruno e il pensiero del Rinascimento.* Firenze: Vallechi, 1925.

GRAF, ARTURO. "Petrarchismo ed anti-Petrarchismo", in *Attraverso il cinquecento*, pp. 3-86. Torino: E. Loescher, 1888.

GREENBERG, SIDNEY. *The Infinite in Giordano Bruno.* New York: King's Crown Press, 1950.

HOROWITZ, IRVING, LOUIS. *The Renaissance Philosophy of Giordano Bruno.* New York: Coleman-Ross Company, Inc., 1952.

KOYRÉ, ALEXANDER. *From the Closed World to the Infinite Universe.* Baltimore: Johns Hopkins Press, 1957.

KRISTELLER, PAUL OSKAR. *The Philosophy of Marsilio Ficino.* Translated by Virginia Conant. New York: Columbia University Press, 1943.

LIPARI, ANGELO. *The Dolce stil nuovo according to Lorenzo de Medici.* New Haven: Yale University Press, 1936.

LOVEJOY, ARTHUR. *Bruno and Spinoza.* Arthur Lovejoy. Berkeley. University of California Press, 1904.

McINTYRE, J. LEWIS. *Giordano Bruno.* London, 1903.

NELSON, JOHN CHARLES. *Giordano Bruno's Gli Eroici Furori and Renaissance Love Theory.* New York: Columbia University Press, 1958.

SALVESTRINI, VIRGILIO. *Bibliografia delle opere di Giordano Bruno e degli scritti.* Pisa: V. Salvestrini Libraio, 1926.

SARAUW, JULIE. *Der Einflus des Plotins auf Giordano Brunos Degli Eroici Furori.* Leipzig: Borna, 1916.

SINGLETON, CHARLES S. *An Essay on the Vita nuova.* Harvard University Press: 1949.

SINGER, DOROTHEA. *Bruno His Life and Thought.* New York: Schuman, 1950.

SPAMPANATO, VINCENZO. *Documenti della Vita di Giordano Bruno.* Firenze, 1933.

TOCCO, FELICE. *Le opere latine di Giordano Bruno esposte e confrontate con le italiane.* Firenze: 1889.

———. *Le opere inedite di Giordano Bruno.* Napoli: 1891.

———. "Le fonti piu recenti della filosofia del Bruno", in *Riconditi della Reale Academia dei Lincei*, I, Ser. V (1892), pp. 503-581.

VALENCY, MAURICE. *In Praise of Love.* New York: Macmillan, 1958.

VOSSLER, KARL. *Die philosophiscen Grundlagen zum süssen neuen Stil.* Heidelberg: Winter, 1904.

Periodicals

CORSANO, ANTONIO. "Il pensiero di Giordano Bruno nel suo svolgimento storico", *Biblioteca storica del Rinascimento,* nuova serie I (1940), pp. 216-222.

LOWES, JOHN L. "The Lover's Maladye of Hereos", *Modern Philology,* XI, (1914), pp. 491-546.

MEMMO, PAUL E., Jr. "Giordano Bruno's *De gli eroici furori* and the Emblematic Tradition", *Romanic Review,* LV : 1 (Feb., 1964), pp. 1-15.

SARNO, ANTONIO. "La genesi *degli eroici furori* di Giordano Bruno", *Giornale critico della filosofia italiana,* II (1920), 158-172.

www.ingramcontent.com/pod-product-compliance
Lightning Source LLC
Chambersburg PA
CBHW030645270326
41929CB00007B/211